alzheimer — s/

Does one lengthen ~~their~~ his life
b/c he has a positive
outlook or does he have
a positive outlook b/c
he has had a healthy full
life?   cause — ~~the~~

Praise for *Breaking the Age Code*

"This exciting new book gives all of us who are aging an opportunity to find ways to positively influence our destiny. Readers can use insights from this book to increase opportunities to shape a better and longer life. The eternal legacy of Maggie Kuhn, the founder of Gray Panthers and one of the most important and effective activists of the twentieth century, is richly demonstrated in this book."

—Jack Kupferman, president, Gray Panthers

"Becca Levy has done a masterful job of describing the importance of aging beliefs on health and well-being at both the individual and societal level. Perhaps most importantly, this book provides practical suggestions on how to maximize the power of positive age beliefs, which can translate into tangible health benefits. An essential read for anyone interested in how we age and how each of us can benefit from adopting positive age beliefs in everyday life."

—Cary Reid, MD, Irving Sherwood Wright
Professor of Medicine, Weill Cornell Medical Center

"At last, Professor Becca Levy shows how we can harness the power of the mind to live a longer and more fulfilling life. She brings a unique perspective about a question we are all concerned with: What happens as we age and get older? She brilliantly shows how we can successfully age. The book offers great insights and it is a must-read!"

—Dr. Itiel Dror, senior neurocognitive researcher,
University College London

"Ageism robs us of optimism for the longer lives we have created, and it disables the future of the young and the old. Becca Levy shows us how this happens, and how we solve it. We need to collectively follow her prescriptions. When we do so, we will create the potential of longer lives that young people can look forward to and older people can live their fullest in."

—Linda P. Fried, MD, MPH, dean,
Columbia University Mailman School of Public Health

"Dr. Levy is a pioneering psychologist and gerontologist. Her wonderful book will inspire us with its solid scientific discoveries and practical advice for longevity. I believe *Breaking the Age Code* will greatly benefit the public!"

—Xin Zhang, PhD, associate professor, Peking University

"This book will shatter some of your basic assumptions about aging—and how we can lead longer, healthier, and happier lives. Becca Levy is the world's foremost expert on the psychology of aging, and she shares rigorous, remarkable evidence that one of the best ways to stay mentally and physically fit is to rethink your stereotypes about what it means to be an older person."

—Adam Grant, #1 *New York Times* bestselling author

"*Breaking the Age Code* is a revolutionary paradigm shift in how we think about aging. Dr. Becca Levy has pioneered a new field of research that reveals how our mindset and beliefs shape our behaviors, our ability to heal, and our lifespan, in invisible but powerful ways. Through cutting-edge science and memorable stories, she shares a new view of aging that will change how we age. Fascinating, inspiring, and moving, this book holds one of the precious keys to a healthy aging society."

—Elissa Epel, PhD, University of California professor
and *New York Times* bestselling author

"Before reading this book, I knew that Becca was a pioneering scientist at Yale. I now know that she is a world-class writer, storyteller, and humanist. Her book is exciting, relevant, and holds the potential to exert powerful global influence on how we age.

This is an extremely profound and timely message that needs to be heard."
　—Sharon Inouye, MD, MPH, Harvard Medical School professor

"Becca Levy is already recognized as one of the world's most respected experts on aging and longevity. Her thought leadership and pioneering research will change many more hearts and minds. Levy's book is a must-read—an urgent and uplifting call to action and road map for a future of longer, healthier, and better lives."
　　　　—Paul Irving, chair of the Center for the Future of Aging

"In this groundbreaking book, filled with stunning scientific insights, captivating stories, and easy-to-use tools, Dr. Levy offers a new way to approach aging and longevity that can benefit readers of any age."
　—James C. Appleby, CEO, the Gerontological Society of America

"Levy has produced a manifesto to inspire us to fight against the scourge of ageism and its negative effects on older adults, and our society. The book is a call to action and provides practical and proven methods to help older adults develop more positive views of their lives, and to inspire all of us to stand up against ageism in our personal life, workplace, and on social media. This book will be remembered as a turning point in the fight against ageism."
　　　　　　—William E. Haley, PhD, chair, Committee on Aging,
　　　　　　　　　　　　American Psychological Association

Breaking
the Age Code

HOW YOUR BELIEFS ABOUT
AGING DETERMINE HOW
LONG & WELL YOU LIVE

BECCA LEVY, PhD

WILLIAM MORROW

An Imprint of HarperCollins*Publishers*

FIRST EDITION

Art credits: Figure 1 (p. 49) from: Levy, B. R., Pilver C., Chung, P. H., & Slade, M. D. (2014). Subliminal strengthening: Improving older individuals' physical function over time with an implicit-age-stereotype intervention. *Psychological Science, 25*, 2127–2135. Reprinted with permission.

Figure 2 (p. 64) from: Levy, B. R., Slade, M. D., Pietrzak, R. H., & Ferrucci L. (2018). Positive age beliefs protect against dementia even among elders with high-risk gene. *PLOS ONE, 13*, e191004. Reprinted with permission.

Figure 3 (page 77) from: Levy, B. R., Moffat, S., Resnick, S. M., Slade, M. D., & Ferrucci, L. (2016). Buffer against cumulative stress: Positive age self-stereotypes predict lower cortisol across 30 Years. *GeroPsych: The Journal of Gerontopsychology and Geriatric Psychiatry, 29*, 141–146. Reprinted with permission.

Figure 4 (p. 93) from: Levy, B. R., Slade, M. D., Kunkel, S. R., & Kasl, S. V. (2002). Longevity increased by positive self-perceptions of aging. *Journal of Personality and Social Psychology, 83*, 261–270. Reprinted with permission.

Figure 5 (p. 138) from: Levy, B. R., Zonderman, A. B., Slade, M. D., & Ferrucci, L. (2009). Age stereotypes held earlier in life predict cardiovascular events in later life. *Psychological Science, 20*, 296–298. Reprinted with permission.

Appendix 1 (p. 205): This figure was created by graphic artist David Provolo.

Library of Congress Cataloging-in-Publication Data has been applied for.

ISBN 978-0-06-305319-9

22 23 24 25 26   FRI   10 9 8 7 6 5 4 3 2 1

To my parents and heroes, Charles and Elinor

CONTENTS

Ideas Bouncing Between the US and Japan

Halfway through graduate school, I was lucky to win a National Science Foundation fellowship to live in Japan for a semester. My goal was to investigate how people aged and thought differently about aging in Japan. I knew that Japanese people had the longest life spans in the world.[1] Although many researchers chalked this up to a healthy diet or genetic differences, I wondered if there might also be a psychological dimension that gave them an advantage.

Before uprooting myself to Japan for six months, I visited my grandma Horty in Florida. As soon as I stepped off the plane, she took one look at me and said, "You need vitamins." She was convinced that graduate school and the dreary Boston weather had run me down, so off we went to buy her version of vitamins—all the oranges and grapefruits that we could carry out of the grocery store.

Grandma Horty was a competitive golfer and, as a former New Yorker, an avid walker, so it was no small feat to keep up with her as she walked with purposeful strides through the store—until she tumbled to the floor. Rushing over, I helped her up and was horrified to spot a bloody gash in her leg.

"Doesn't hurt," she reassured me, between clamped teeth. She even forced a smile, ever the stoic. "You should see the other guy," she joked.

The "other guy" was down at our feet: a wooden crate with reinforced corners of sharp, jagged metal; one corner was now dripping blood. We left our baskets and I helped my grandmother gather up the contents of her handbag, which had scattered across the floor.

On the way out, she confronted the owner, who glanced up for a second when he heard her fall, before returning to the tabloid he'd been leafing through at the counter.

"You shouldn't leave crates in the middle of your store," my grandmother told him, much more politely than he deserved. "I could have hurt myself." Blood was dripping down her calf.

The owner looked her over and then peered at the crate in the middle of the aisle. "Well, maybe you shouldn't be walking around," he said icily. "It's not my fault old people fall down all the time. So don't go around blaming me."

Horty's jaw practically dropped to the floor. As for me, I felt like swiping his tabloid off the counter, but I just glared at him and ushered my grandmother into the car. Over Horty's objections, I took her straight to the doctor. Her leg turned out to be fine—a dramatic-looking but superficial cut, the doctor said. He added that she seemed in fact quite healthy.

I thought that would be the end of it, but some profound change had taken place that afternoon. That night, Horty asked me to water her avocado tree, which she normally loved doing herself. The following day, she told me she didn't trust herself to drive and asked me to take her to a hair appointment. She seemed to be reliving the grocery owner's words and questioning her competency as an older person in a way she never had before.

Fortunately, by the time I flew to Japan, Horty had pulled out of her ageism-induced funk. The morning before my departure, she insisted on taking me for a brisk, long walk, to stretch my legs before the long plane ride. When we returned, she handed me a handwritten list of restaurant recommendations, from a visit to Japan with my grandfather, two decades earlier.

But as I waved goodbye to Horty and headed off to Tokyo, I couldn't help but wonder: If a few negative words could affect someone as strong and spunky as Horty, what were negative age stereotypes doing to us as a country? What power did they have to actually change the way we age? And what power could *we* have if we changed the way we thought and talked about aging?

Aging Rock Stars in Tokyo

As I settled into my new life in Tokyo, my mind would often fly back to Grandma Horty, watering her avocado tree in the cooling twilight of a Florida evening, and I wondered what she would make of this place where centenarian sushi chefs were constantly feted on television, and older relatives were served first at meals.

I was still in Japan when a national holiday called *Keiro No Hi* rolled around. Walking through Shinjuku Park that day, I passed crowds of weight lifters, some of them shirtless, some in skintight leotards, all of them in their seventies and eighties, strutting around, lifting weights, flaunting their well-toned bodies. All over the country on this holiday, people were crisscrossing the archipelago by high-speed train, boat, and car as they returned home to visit their elders. That day, restaurants would serve free meals to seniors; for those less mobile, schoolchildren would prepare and deliver bento boxes full of fresh sushi and delicately fried tempura.

Keiro No Hi translates to "Respect for the Aged Day," but evidently the Japanese were already doing this every day of the year. Music classes were full of seniors trying out the electric slide guitar for the first time at age seventy-five. Newsstands were lined with colorful manga, the popular comic books for readers of all ages, telling stories of older people falling in love. The Japanese treated old age as something to enjoy, a fact of being alive, rather than something to fear or resent.

In the US, it was a different cultural picture. It wasn't just my grandmother's interaction with the ageist store owner; it was everywhere: the billboards for "age-defying" skin treatments, the late-night ads for local plastic surgeons going on about wrinkles like they were generals describing hostile enemy forces, the infantilizing greetings older people endured in restaurants and movie theaters. Everywhere I looked, in TV shows, in fairy tales, and online, old age was treated as though it meant forgetfulness, weakness, and decline.

In Japan, it became clear to me that the culture we're in impacts how we age. Take menopause, for instance. I learned that Japanese culture doesn't typically make a lot of fuss around it, treating it as a natural part of aging that can lead to a valued phase of life, rather than as fodder for those Western stereotypes of female irritability and sexual obsolescence that characterize menopause as a midlife affliction. And the result of the Japanese being less likely to stigmatize this natural aspect of aging than their peers in North America? Older Japanese women are much less likely to experience hot flashes, as well as other symptoms of menopause, than women of the same age in the US and Canada.[2] And Japanese older men, who are treated, culturally, "like rock stars in their country," according to the anthropologist who led this study, were found to have higher testosterone levels than their European counterparts.[3] This suggests that your libido ages differently depending on the way your culture perceives and treats aging.

I began to wonder just how much culture impacts individual age beliefs—how we think about older people and getting older. And I was curious about the extent to which these individual conceptions, in turn, influence the aging process. Could their age beliefs help to explain why the Japanese have the longest life span in the world?

I had gone to graduate school to study social psychology, the science of how individuals' thinking, behavior, and health are impacted by their society and the groups they belong to and interact with. I wanted to focus on the experience of older people, who were being left out of most psychology studies. The puzzle in front of me now was how to measure the impact of something as amorphous as culture on something as definite as our biology.

Impact of Age Beliefs on Aging Health

When I returned to Boston, I set about testing the impact of cultural age stereotypes on the health and lives of older people. In study after study I conducted, I found that older people with more-positive perceptions of aging performed better physically and cognitively than those with more-negative perceptions; they were more likely to recover from severe disability, they remembered better, they walked faster, and they even lived longer.[4]

I was also able to show that many of the cognitive and physiological challenges we think of as linked to growing old—things like hearing loss and cardiovascular disease—are also the products of age beliefs absorbed from our social surroundings. I found that age beliefs can even act as a buffer against developing dementia in people who carry the dreaded Alzheimer's gene, *APOE* ε4.

This book is about how we think about aging and how these beliefs impact our health in ways big and small. It is for anyone who hopes to age well. In the following pages, I will also explore the anatomy of negative age stereotypes—how they build up in us, how they operate, and how they can be changed. Although these stereotypes are developed within cultures over hundreds of years, and assimilated across the lifetimes of individuals, they are in fact quite brittle: they can be chipped at, shifted, and remade.

In my Yale lab, I have been able to improve people's memory performance, gait, balance, speed, and even will to live by activating positive age stereotypes for just ten minutes or so. In this book, I will show you how priming, or the activation of age stereotypes without awareness, works, what it says about the unconscious nature of our stereotypes, and how we can strengthen our ideas about aging.

With the right mindset and tools, we can change our age beliefs. But to get at the origin of the beliefs, the ageist culture needs to change. To better understand how we got here and what is possible, we'll look at cultural alternatives around the world and throughout history. We'll look at stories of successful aging and visit the homes, memories, and outlooks of athletes, poets, and activists, movie stars, artists, and musicians. We'll look at what it takes to change our culture and learn how better integrating older people into our communities can lead to greater collective health and wealth.

Demographically, we are at a crossroads. For the first time in human history, there are now more people in the world over the age of sixty-four than under the age of five.[5] Some politicians, economists, and journalists are wringing their hands over what they call "the silver tsunami," but they're missing the point. The fact that so many people are getting to experience old age, and doing so in better health, is one of society's greatest achievements. It's also an extraordinary opportunity to rethink what it means to grow old.

When Grandma Horty died, many years after the ageist grocery store encounter, my family and I got together to celebrate her life, which had been both ordinary and remarkable. She lived through most of the twentieth century, witnessing both its progress and its atrocities.

She lived her life to the fullest, even after raising my dad and after caring for my grandfather, as he dwindled away under the strange, dreadful weight of Alzheimer's. In her eighties and nineties, she traveled and played golf and went on long walks with her friends. She threw elaborate costume parties and wrote us letters that reflected her unforgettable personality, at once assertive and witty.

When I received a letter from her, it was as if she was sitting right there in the room with me. In the last decade of her life, she and my father would often turn on the TV to watch the same headline-grabbing court trial while they talked on the phone about it. Though my father was in frigid New England, and Grandma Horty in sunny, damp Florida, it was as if they were sitting together in the living room, gossiping about which lawyer was currying favor with the judge, and whose tie looked the goofiest.

Grandma Horty never let you forget that she was there, so her absence rang loudly when we went through her things to find new homes for them after she died. Her basement, especially, was a testament to the richness of the life she lived. Before she began caring for my grandfather, Grandma Horty and Grandpa Ed spent decades visiting far-flung places, collecting souvenirs. I thought that her basement was a cave of wonders. There were old French perfume bottles, wonderful flowing scarves from Italy and Indonesia, delicate little carved boxes from Morocco. And amid it all, a small, Japanese-style folding screen. The panels were made of shoji rice paper and painted over with a cherry tree in bloom. On the screen was a Post-it with my name on it. I remembered Grandma Horty once telling me, after I admitted to her that it sometimes took me a while to focus when I sat down to write, that Sigmund Freud used a screen to sequester himself from the world and summon his concentration. I was touched that she had left me this screen, and I thought back to my first visit to Japan, and to the episode of her fall.

Shortly after my grandmother's death, I made a surprising discovery. While analyzing data from my study about the lives and outlooks of the inhabitants of the small town of Oxford, Ohio, I found out that the single most important factor in determining the longevity of these inhabitants—more important than gender, income, social background, loneliness, or functional health—was how people thought about and approached the idea of old age.[6] Age beliefs, it turns out, can steal or add nearly eight years to your life. In other words, these beliefs don't just live in our heads. For better or worse, those mental images that are the product of our cultural diets, whether it's the shows we watch, the things we read, or the jokes we laugh at, become scripts we end up acting out.

When I first landed on this longevity finding, I thought of Grandma Horty and of how lucky my family was to have had her with us until her death at ninety-two. I thought of how fortunate she was to have approached aging the way she had. I thought of the gift of those extra years, and where it came from. Did we have seven or eight more years with Horty because she embraced life in those later years? If there is a code to aging well—a system or a method—age beliefs are a part of it.

Our lives are the product of so many different factors that we can't control: where we are born, and to whom, what's in our genes, and which accidents befall us. I am interested in identifying those factors we *can* control to improve our aging experiences and health. One of those factors is the way we think about aging and conceptualize the life cycle. That's what this book is all about: how we can, as individuals and a society, change how we think of ourselves and those around us as we grow old, to enjoy the benefits of this change.

1

The Pictures in Our Head

Each fall, I start my Health and Aging class at Yale by asking my students to think of an old person and list the first five words or phrases that come to mind. It can be someone real or imagined. "Don't think about it too much," I tell everyone. "There are no right or wrong answers. Just write down whatever associations pop into your head."

Try it now. Think of the first five words or phrases that come to mind when you imagine an old person. Write them down.

Once you're done, take a look at your list. How many of these are positive? How many are negative?

If you are like most people, chances are your list includes at least a few negatives. Take this response from Ron, a seventy-nine-year-old violin maker outside of Boston: "Senile, slow, sick, grumpy, and stubborn." Now, consider this description from an eighty-two-year-old woman named Biyu in China who was in her former workplace, a pencil factory, to pick up her pension check: "Wise, loves Peking Opera, reads to grandchildren, walks a lot, and kind."

These two clashing visions reflect the vast range of age beliefs that predominate in different cultures—beliefs that determine how we act toward our older relatives, organize our living spaces, distribute health care, and form our communities. Ultimately, these beliefs can also determine how older people think about themselves as well as how well they hear and remember, and how long they live.

Most people don't realize they hold preconceptions about aging, yet everyone, everywhere, does. Unfortunately, most of the world's prevailing cultural age beliefs today are negative.[1] By examining these beliefs and discovering their origin, together with how they operate, we'll have a basis for changing not only the narrative of aging, but the very manner in which we age.

What Are Age Beliefs?

Age beliefs are mental maps of how we expect older people to behave based on age. These mental maps, which often include pictures in our heads, become activated when we notice members of the group in question.

When I talk about an "older person," by the way, I'm usually referring to someone who is at least in their fifties, but there really is no set age threshold. How "old" we feel, rather than the number of years under our belt, is often based on cultural cues, such as becoming eligible for "senior discounts" or Social Security, or being nudged into retirement. There is actually no single biological marker to identify when someone has reached old age, which means that old age is a somewhat fluid social construct. This is one of the reasons age beliefs, with their associated expectations, are so powerful: they define *how* we experience our later years.

Expectations can be quite useful in many situations. When we come across a closed door, we can expect, based on previous experiences, that it will either be locked or unlocked. We generally don't have to ask ourselves whether the door will fall down flat or burst into flames if we give the handle a wiggle. We can thank our brains for this ability to process situations quickly, visually, and often automatically, which is why there's no need to relearn how a door works. Instead, we can rely on what we already know to be familiar. This is pretty much how we get through the world every single day: by generating and then relying on expectations.

Age beliefs are expectations about people, not doors, of course, but they operate in a similar way. Like most stereotypes or mental shortcuts, they are the product of natural, internal processes that begin when we are babies as a way of sorting and simplifying the overwhelming amount of stimuli in the world. But they are also the products of external societal sources, such as schooling, movies, or social media and the ageism that operates in these realms.

Linking Structural and Implicit Ageism

Stereotyping often happens unconsciously. Our brains make decisions up to ten seconds before we're aware of it.[2] Nobel Prize–winning neuroscientist Eric Kandel found that about 80 percent of our mind works unconsciously.[3] This is all fine and good when you're reaching for the door handle, but when forming impressions or making decisions about people, it's a different story.

Stereotypes are the often-unconscious devices we use to rapidly assess our fellow human beings. A lot of the time, however, these pictures aren't based on observed or lived experiences, but rather, are absorbed uncritically from the external social world.

Most of us like to consider ourselves as capable of thinking fairly accurately about other people. But the truth is, we are social beings who carry around unconscious social beliefs that are so deeply rooted in our minds that we don't usually realize they've got their hooks in us. This can result in an unconscious process called "implicit bias," which automatically influences us to like or dislike certain groups of people. Implicit bias is hard to mitigate or even just to accept, because it so often goes against the grain of what we consciously believe. A further complication is that implicit bias often reflects structural bias.

Structural bias refers to policies or practices of societal institutions, such as corporations that discriminate against workers or hospitals that discriminate against patients. It is frequently intertwined with implicit bias. For within institutions, the discrimination may operate without the managers' or doctors' awareness and therefore can be considered implicit. But at the same time, it is often structural insofar as the discrimination reinforces the power of those in authority while withholding power from those who are marginalized.

To examine both types of bias, researchers asked fellow scientists, people who usually think of themselves as objective and fairminded, to evaluate résumés of male and female applicants for a job. In almost every case, male applicants were more likely to be hired and offered much higher salaries than their female counterparts, in spite of what turned out to be résumés that were identical in every way, other than the use of traditionally masculine or traditionally feminine names.[4] There exists a similar culture-based racial bias: studies show that job seekers who added typical "white" identifiers to their résumés received significantly more calls for interviews than those without these identifiers.[5]

The same structural and implicit bias, or ageism, exists toward older job applicants. A study found that when résumés are otherwise identical, employers tend to offer the job to younger applicants.[6] This hiring pattern occurs again and again, despite abundant research showing that older workers are usually more reliable and skilled than younger workers.[7] Similarly, when doctors are given identical case studies of patients with the same symptoms and likelihood of recovery, these doctors tend to be far less likely to recommend treatments for the older patients compared to the younger patients.[8]

The line between structural and implicit bias is thin and quite porous. Culture-based structural biases seep into our beliefs, where they are then often activated without our awareness. As a result, multiple studies show that all of us, no matter what we consciously believe, have unconscious biases.

Unconscious Running

As someone who studies self-stereotypes for a living, I didn't think I was susceptible; but of course, there's a difference between what you *think* you know and what you *actually* know. There are those moments in life that reveal the uncomfortable gap between the two.

Last year, I decided to run a 5K to benefit a charity that a friend of mine is involved with. It was a chilly fall Sunday morning, and my bed was particularly warm and cozy, so I hit the snooze button one time too many and arrived late. I barely had time to pin on a race bib and lace up my sneakers when the starting pistol went off. I was about two hundred yards into the race, running past a cluster of tall elm trees, when I heard a nasty popping sound and the back of my knee suddenly flooded with pain. I stumbled a bit and groaned. An image immediately appeared in my mind—a scene from the sci-fi film *Lucy*, in which Scarlett Johansson plays a woman whose body parts quickly disintegrate into sparkly dots after smugglers implant a dangerous drug in her stomach. As I envisioned it, my body, my once trusted and reliable friend, was precipitously disintegrating in the same fashion, only age, and not some sci-fi lab drug, was the culprit.

I hobbled past the finish line and even gave a grim smile to the friend who had encouraged me to participate. After I drove home, I limped inside, complaining that my middle-aged body was succumbing—all too early—to the ravages of age. Here I was, I thought, facing the sad and premature end of my running days.

Then, my husband, who is a doctor, inspected my leg and told me that I had badly pulled a muscle.

At that moment, my teenage daughter chimed in. She'd been on her laptop that morning and had watched me rush out the door for the race.

"You got there late, didn't you?" she asked.

I nodded.

"Did you warm up?"

I shook my head. Who has time to warm up when you show up late?

She smiled. "Well, there you go."

We all like to run in my family. We all know that warm-ups activate your muscles, stretching and lengthening them to protect them from being pulled too far and tearing. My other daughter, less than a month prior, had pulled a leg muscle of her own by dashing off on a run without stretching.

There you go.

Instead of being relieved that my body wasn't suddenly falling to pieces, I was troubled. I had instinctively attributed my injury to something other than skipping a warm-up. Instead, I had blamed my age: my mind had made connections that I don't consciously believe—that your body falls apart as you age. And I have been studying aging since the start of graduate school. More than most, I should know that this isn't the case. So what happened? The negative stereotypes I had absorbed since childhood from the surrounding culture materialized in a sudden fear of age-related frailty, which led me to misattribute the cause of the pain in my knee.

This is one of the most harmful things about negative age stereotypes: they don't only color our actions and judgments toward other people; often, they influence how we think about ourselves, and these thoughts—if they are not counteracted—can impact how we feel and act.

When I was starting out as a social psychologist, the existing studies on age stereotypes were limited to how age beliefs influenced the views and behavior of children and young adults toward older people. The research ignored the ways in which age stereotypes impacted older people themselves. But after watching my grandmother absorb and respond to the negative age stereotype directed at her from the ageist store owner, I became convinced that to reduce the likelihood that this kind of event would occur in the future and to find ways to harness the power of age beliefs to bring about benefits, I would first need to understand how our beliefs about older people affect the way we ourselves age.

How Our Culture's Beliefs Become Our Own

To better understand how culture-based age stereotypes get under our skin, I developed a framework called stereotype embodiment theory (SET), which proposes that negative age beliefs bring about detrimental health effects that are often, and misleadingly, characterized as the inevitable consequences of aging. At the same time, positive age beliefs do the exact opposite: they benefit our health.[9] My research, on which I based these twin concepts, has been confirmed by over four hundred studies conducted by other scientists in five continents.[10]

According to SET, there are four mechanisms involved in how age stereotypes affect our health. They

1. are internalized from society starting in childhood and continuing throughout the life span;
2. operate unconsciously;
3. increase in power as they become more self-relevant; and
4. impact health through psychological, biological, and behavioral pathways.

Now I will illustrate how age beliefs draw on these interlocking mechanisms to slip under our skin and impact the age code over the course of our lives.

SET Mechanism 1: Internalization Across the Life Span

Even though we often assume kids aren't tainted by the negative beliefs of adults, children as young as three have already internalized the stereotypes, including age stereotypes, of their culture enough to express them.[11] A study of American and Canadian youngsters found that many of them already viewed older people as slow and confused.[12] It turns out the tendency to categorize starts even earlier: infants as young as four months distinguish and sort faces by age.[13]

While we absorb all sorts of negative stereotypes from our culture and society, we're particularly susceptible when it comes to negative age beliefs. There are four reasons that many absorb these ideas like a sponge in a bowl of water. First, their sheer prevalence; according to the World Health Organization, ageism is the most widespread and socially accepted prejudice today.[14] Second, unlike race and sex stereotypes, we encounter age stereotypes decades before they refer to our own age group, so we rarely question or try to resist them. Third, society often segregates older people in terms of where they live, work, and socialize; children, who notice the ways that older persons are segregated, infer that these social divisions are caused by meaningful, inherent differences between age groups rather than what it often really is: people in power marginalizing older people.[15] Fourth, these stereotypes are then frequently reinforced over our lifetimes as we are bombarded by messages in advertisements and media about older people.

SET Mechanism 2: Unconscious Operation

As the psychoanalyst Carl Jung observed: "Until you make the unconscious conscious, it will direct your life and you will call it fate." A reason age stereotypes are so effective at impacting our health is that they often operate without our awareness.

To examine how the numerous ageist tropes that permeate our culture often influence us, I found that an effective way to activate age stereotypes is to present them subliminally. In our experimental studies, people sit in front of computer screens as words flash by at speeds that are so fast that they are either unseen or else seen as a blur. This allows for perception without awareness. These flashed words are either positive stereotypes, such as "wise," or negative ones, such as "decrepit." Afterward, participants perform a range of simple tasks, such as strolling down a hall. In this way, I've demonstrated that age stereotypes unconsciously affect everything from how neat or messy our handwriting is to how quickly we walk.[16]

SET Mechanism 3: Self-Relevance of Age Stereotypes

The worst effects of negative age beliefs don't hit us until we get older because of the self-relevance that these beliefs then acquire. If you misplace your car keys at age twenty-five, you probably won't think too much of it. If the same thing happens when you are seventy-five, you might worry about impending senility since you've spent most of your lifetime absorbing stereotypes about people over sixty becoming mentally incompetent—even though this stereotype is not accurate, as we'll explore in the next chapter.

Pretend for a moment that you are an older person who, when you were a child, heard your parents complain about *their* parents being absentminded and blamed it on their advancing age. As you entered your twenties, this belief was likely reinforced by similar messages about aging in ads, movies, and books. By the time you entered middle age, you began to mentally label instances of other people's forgetfulness as something to do with getting old. And as you finally enter your own old age, whenever you can't recall something, you blame it on aging. When you do this, you're actively manifesting the stereotype you grew up hearing applied to older people, but now you're directing it at yourself. This, in turn, can lead to stress, which can reduce memory performance.[17] After a lifetime occupying the inner reaches of your mind, in later life the destructive age stereotype is likely to take its toll (unless it is counteracted by some of the strategies presented later in the book).

SET Mechanism 4: Three Pathways Age Beliefs Follow to Get Under Our Skin

There are three pathways age beliefs use to act on health outcomes: psychological, behavioral, and biological. An example of the *psychological* pathway is the low self-esteem that develops among elderly people who have assimilated negative age beliefs.[18] A letter I recently received from an older Englishwoman states in its opening: "Frankly I feel ashamed to be old. Why? Because society tells me it is shameful."

The *behavioral* pathway plays out as older people take in negative age beliefs and develop fatalistic attitudes about the inevitability of declining health in later life. They'll sometimes then cut back on healthy behaviors, which in this grim light appear to be pointless. My team found that older people with negative age beliefs take fewer prescribed medications and exercise less.[19] The situation can become a self-fulfilling prophecy: negative beliefs about aging lead to unhealthy behaviors that worsen health, which then reinforces the original negative age beliefs.

The third pathway is *biological*. We've found that negative age beliefs can increase biological markers of stress, including the hormone cortisol and a substance in our blood called C-reactive protein (CRP).[20] Over time, more frequent and higher spikes in stress biomarkers can lead to earlier death.[21]

And there you have it. The misfortune trifecta. That's how negative age beliefs seep from our culture into our health, shortening our life spans and diminishing our sense of well-being. I know this sounds bad, and it is. But these physical manifestations are not inevitable. Negative age beliefs can be resisted and reversed, leading to positive psychological, behavioral, and biological outcomes.

From Targeter to Targeted: The Toxicity of Age Stereotypes

It may be hard to accept that age stereotypes can have such profound effects on the ways we act and feel, but it's not just age beliefs: other kinds of self-stereotypes can have an impact. For instance, studies show that Black test takers tend to perform worse than white test takers when asked to identify their race on a demographic questionnaire prior to the test.[22] When the demographic questionnaire was omitted, however, there was no significant difference in test scores. The stereotype linking race with intellectual performance was so strong that all it required to take hold was for test takers to be asked to identify their race.

Another study revealed that when female participants were shown actual TV commercials that played up gender stereotypes, they often avoided leadership roles in subsequent tasks doled out in the experiment, even though the commercials had nothing to do with leadership.[23] One commercial showed a young woman who becomes so excited about being given a new beauty product that she bounces on her bed with joy. Activating just one stereotype about a group, in this instance, the stereotype of women as focused on appearance, can open the floodgates to a slew of other stereotypes and associations, one of which is that women are not good leaders.

Also heavily featured in commercials are age stereotypes, which operate differently than racial or gender stereotypes, since they do not become self-relevant until later life. This process is made more complicated by the fact that older people, before reaching old age, were the targeters, viewing older people in stereotypical ways, rather than the targeted, and had no need to question or resist these stereotypes themselves. So when they reach old age, people often still identify with the young, rather than their new group, the old.[24]

It is important to note that age beliefs are not the same thing as pessimistic or optimistic thinking. You might think positive age beliefs are just one facet of positive thinking, and negative age beliefs a form of negative thinking. But in my research, I have found that above and beyond general emotions, such as happiness or gloominess, *age beliefs* are what drive outcomes, including how well we recall information, or how quickly we walk around the block.[25] That is, it's age beliefs, above and beyond the emotional outlooks of whether, say, you are a glass half-full or half-empty kind of person, that harm or improve our health.

A Two-Thousand-Year-Old Man with Positive Age Beliefs

Most research conducted on beliefs about groups regardless of whether they are based on race, gender, ethnicity, or age has focused on negative beliefs. In contrast, I have long been curious about the potential benefit of positive beliefs.

Take the filmmakers Carl Reiner, creator of *The Dick Van Dyke Show* and director of the popular Steve Martin film *The Jerk*, and Mel Brooks, who wrote and directed *Young Frankenstein* and *The*

Producers. Reiner died at age ninety-eight, survived by his lifelong friend Brooks, now ninety-five. Both drew on their positive age beliefs as they got older and surpassed the average male life span by at least fifteen years. They were extremely productive throughout their nineties, with Reiner writing five books and Brooks continuing to act, write, and produce. They were also quite happy. Their friendship grew closer over the decades. Until Reiner died, the two of them would have dinner and watch *Jeopardy!* every night at Reiner's house, following it up with an amusing movie, sometimes one of their own.[26]

At ninety-five, Reiner starred in an HBO documentary about life as a nonagenarian, called *If You're Not in the Obit, Eat Breakfast.* The people he interviews (which include Brooks) are funny, self-deprecating, and happy. They describe their lives as productive and full of meaning and complain about the condescension they face from society for being old. Norman Lear, who created many of the most memorable sitcoms of the 1970s (including *All in the Family, Maude,* and *The Jeffersons*), tells Reiner in the documentary: "Because I'm ninety-three, I'm supposed to behave a certain way. The fact that I can touch my toes shouldn't be so amazing to people." He is now ninety-nine and just created a new sitcom with an all-Latino cast.

Mel Brooks and Carl Reiner started to express their positive age beliefs many decades earlier. When I was growing up, my parents used to play old recordings of their famous comedy bit "The 2,000-Year-Old Man," which featured Reiner as the interviewer and Brooks as the 2,000-year-old man, improvising stories and jokes in a thick, bemused Yiddish accent. The humor of it came from the eminently quotable, perfectly timed, Borscht Belt zingers of this very ancient man.

Although Reiner and Brooks created the skit to amuse themselves and their friends at parties when they were in their early thirties, their comedy routine also reflected and promoted their positive images of aging. Many of the 2,000-year-old man's humorous observations have to do with experience-based skills that allow him to survive in a chaotic world. In addition, he displays excellent memory, which allowed him to recite the lyrics of what he describes as the first song ever sung, an ancient Aramaic chant that happened to sound just like the popular jazz standard "Sweet Georgia Brown." It is a refreshing contrast to most of the current "aging humor," often featured in stand-up comedy and television shows, which makes fun of older people's minds and bodies.[27]

Changing Age Perceptions

The good news is that we are not born with a set of age beliefs and once we take them in they are not set in stone. We see this first and foremost in how radically age beliefs vary from culture to culture. I found that in China, when I asked people to describe the first words or phrases that came to mind to describe an old person, the most common response was "wisdom," whereas in the US, the first image to come to mind is usually "memory loss."

We also know that age beliefs are malleable because they change throughout history, and I have been able to shift them from negative to positive in research studies.[28] Later in the book, we will explore these cultural differences, as well as the historical and experimental shifts of age beliefs. In addition, based on these patterns, we will present a strategy for improving these beliefs.

The Far Reach of Age Beliefs

Age beliefs affect almost all aspects of our lives including access to health care and work opportunities. Older patients who go to mental health professionals with depression are less likely to be treated adequately, if at all, because it's widely believed by these profession-als that depression is a normal part of aging.[29] It's not just assump-tions about mental health: it's roundly dismissing older patients as not worthy of care.

The geriatrician Louise Aronson shared a story about a meet-ing at a major US hospital where doctors were discussing the com-plicated case of an older patient who had been brought in from a nearby nursing home. Halfway through the meeting, one of the doctors—a department chief—stood up and suggested that he had a solution for this complex case.

"I figured it out. We just need to have nursing homes be a hun-dred miles away from our hospitals."

Everyone laughed. The underlying assumption was that older patients are a waste of time, care, and money. Aronson adds: "If someone had said this about women or people of color or LGBTQ people, there would have been outrage. In this case, there was none. It makes you want to cry."[30]

Negative age beliefs run rampant in the job market, too, where only a third of Americans aged sixty-eight are still employed,[31] in part due to the assumption that older workers are inefficient and need to be cleared out. The median age of employees is twenty-eight at Facebook, thirty at Google, and thirty-one at Apple.[32]

Is it that only young people have the requisite skills to work in tech, or that the field is rampant with negative age beliefs? We ought to ask JK Scheinberg, the legendary Apple engineer who ran the company's top-secret Marklar Project, which moved Apple's operating system to Intel and made the MacBook the roaring suc-cess that it is today. After twenty-one years at Apple, JK decided to retire early, while still in his fifties, but soon found that he was growing bored and restless. Thinking that he knew one easy way to make himself useful, he applied for a part-time job at his local Apple Store's Genius Bar, where he was told by his interviewers that they would be in touch. He never heard from them. JK figured that's because he was the oldest candidate by several decades at his interview.[33]

Negative age beliefs are the most tolerated of all types of implicit bias. They push older people out of vibrant neighborhoods and doctors' clinics and the workforce, for no reason other than their age.

Aging is a biological process, yet it doesn't exist in some strictly biological dimension, independent of our beliefs and practices around what it means to be or grow old. Too often, we don't realize that our age beliefs are the product of cultural biases, rather than scientific facts. We forget that only 25 percent of our health is due to genes.[34] That means three-quarters of our health is determined by environmental factors, many of which we can control. As my research has shown, one of those controllable factors is age beliefs.

In this first part of the book, I'll present scientific findings about our age beliefs, together with stories about a range of people including artists, movie stars, and athletes, to illustrate the ways these beliefs impact our health, biology, memory, and general well-being. Later in the book, I'll share evidence-based strategies to help you harness the formidable power of age beliefs and fight structural ageism in ways that can help you, those you love, and the world we live in.

2

Anatomy of a Senior Moment

Sometimes memory short-circuits. Forgetting the name of the hero in the movie you just saw or walking into a room to grab something only to forget what it was. It's a frustrating mental state, and it happens to everyone. We often refer to these vexing lapses in memory by saying we're having "a senior moment." But why "a senior moment" when it's something that can happen at any age?

The term first appeared in print in 1997, in a piece in the *Rochester Democrat & Chronicle* in which a columnist quoted an older vacationing banker who had forgotten the score of his ongoing tennis match.[1] Since then the expression has infected the lexicon, first in the US and, more recently, beyond it.[2] When I give talks in other countries, I sometimes ask who has heard the term, and just about everyone in the room quickly raises a hand.

The reality is that these "moments" have nothing specifically to do with "seniors" or old age. Brief memory lapses have been around forever. Nearly 150 years ago, William James, the "father of American psychology," described the phenomenon as a gap in the mind "that is intensely active. A sort of wraith of the name is in it, beckoning us in a given direction, making us at moments tingle with the sense of our closeness, and then letting us sink back without the longed-for term."[3] Obviously, elderly people aren't the only ones who get struck with occasional moments of forgetfulness, which is what makes the innocuous or even cute-seeming term "senior moment" such a perfect microcosm of the insidious mechanisms and effects of ageism: it cloaks a complex and malleable process (memory) in pseudoscientific legitimacy, thus packaging a widespread anxiety into a derogatory idea that gets applied to everyone above a certain age.

The truth about memory is that there is tremendous variability among people's brain function as they age. A growing number of studies reveal that neuroplasticity, the brain's ability to stay flexible and sprout new neural connections, which was long thought of as a hallmark of young brains, actually continues throughout the aging process. This suggests that the all too commonly accepted stereotype that brain inevitably deteriorates as we get older is false.[4]

As we will soon explore, in later life some forms of memory improve, among them semantic memory, which is recall of general knowledge, such as the possible colors of an apple; some stay the same, including procedural memory, which is recalling how to carry out a routine behavior, such as riding a bike, and some decline, including episodic memory, which is recall of a specific experience that occurred at a particular time and place, such as seeing lightning flash across the sky above your house during last night's storm.[5] Further, this last type of memory, which is assumed to decline in all older persons, often improves in this group with interventions.[6] Well, still, you might ask, if *some* forms of memory decline in *some* people, *some* of the time, isn't there validity to the "senior moment" term? The reality is that memory lapses can occur at any age, and our brains form new connections in later life that can compensate for these occasional losses.

In short, what causes certain forms of memory to decline isn't necessarily aging itself, but rather the way we approach and think about aging—the way culture tells us, and the way we tell ourselves, how to grow old.

Memory Achievements of Chinese and Deaf Elders

Early in my career, I wondered what role, if any, culture and age beliefs play in memory patterns in later life. As the prevalence of the term "senior moment" suggests, memory decline is one of the most widespread stereotypes of old age in North America and Europe.[7] In one of the first studies I conducted after returning from Japan, I investigated whether stereotypes like this could impact memory performance.[8] I chose to look at three different cultures with different age beliefs: Deaf Americans,[9] hearing Americans, and mainland Chinese.

Why these three particular cultures? I chose Chinese culture for its two thousand years of Confucian values emphasizing filial piety and respect for the old, which have left a significant imprint on contemporary life in China. Today, multigenerational households, often led by older members, are a norm rather than an exception, and elderly people often speak of their advanced years with pride.[10]

As for the American Deaf community, I started to admire that culture in graduate school, when I first learned about its positive age beliefs from a book by anthropologist Gaylene Becker.[11] Old age in the Deaf community is often an extremely social, lively, and interdependent stage of life. Mostly, that's a consequence of how intergenerational the Deaf community tends to be. Over 90 percent of Deaf individuals are born to hearing parents, so when young Deaf people meet older Deaf people, they often develop admiration and form strong ties with these role models with whom they share an identity.[12] As a result, positive age stereotypes abound in the Deaf community, and its older members often feel good about themselves and form a close network with their peers. Becker explains: "During the course of my fieldwork, I saw a pattern in the interaction of the aged Deaf recurring again and again. When individuals were in a group of Deaf people, they were talkative, confident, outgoing and relaxed."[13]

To learn more about Deaf culture, I enrolled in an American Sign Language class offered at a local community center. It was taught by an older Deaf man who signed as if he were choreographing a beautiful dance. One day after class, I worked up the nerve to ask if I could talk to him about his views on aging. By the end of our conversation, he agreed to help me recruit Deaf participants from the intergenerational Boston Deaf Club.

I also recruited old and young hearing American participants from a Boston senior center and a youth organization, and old and young hearing Chinese participants from a Beijing pencil factory where the employees were mostly younger workers, but retired older workers still came around each month to fetch their pension checks.

Looking at participants from all three of these cultures allowed me to eliminate alternative explanations, such as language, for whatever significant patterns in memory I might find. If I compared only American hearing and American Deaf participants, it could be that the Deaf participants had developed a memory advantage from years of signing. If I compared only Chinese hearing and American hearing participants, it could be that the Chinese had better memory because of their exposure to a language relying on pictograms rather than an alphabet. But by including both Deaf American and Chinese participants, we could focus on a unique factor shared by both cultures: widespread positive age beliefs.

I designed the study to test the type of memory that cognitive specialists often claim declines with old age:[14] episodic memory. This is used when you remember a person or object from a specific visual-spatial context, such as a No Hunting sign with a bullet hole in its corner that you noticed driving into a national park.

To measure participants' attitudes toward aging, I started by asking them to respond to the same Images of Aging exercise you may have tried out in the last chapter, which involves naming the first five words or phrases that come to mind when thinking of an older person. Then participants took the Myths of Aging quiz,[15] which consists of twenty-five true-false statements about aging, such as "Depression is more frequent among the elderly than among younger people" and "Old people usually take longer to learn something new." These answers allowed me to gauge participants' bias toward aging. (The above two statements are false, by the way.)

There were some cultural challenges that I enountered. It was unclear to me, for instance, whether certain of the Images of Aging responses from participants in China were positive or negative, once translated into English, since many of these references were culture specific. Among them were "Able to organize the masses" and "Gives remaining heat to society." Fortunately, my assistant, who grew up in China, was able to rate responses in terms of positivity or negativity (the above two responses, it turns out, are quite positive).

Though we weren't testing a new memory drug, the results were just as mind-altering. Of the older participants, the American hearing group expressed the most-negative age beliefs and performed worse on all four memory tasks. The Chinese elders, the group with the most-positive age beliefs, performed best across the board. I was startled to discover that in China, the older participants performed just as well as their younger counterparts. In other words, if you are an elderly Chinese person, you can expect your memory to work basically as well as your grandchildren's. The elderly American Deaf participants, who had more-positive age beliefs than their hearing counterparts, performed much better than the elderly American hearing participants. In contrast to the older participants, the younger participants performed similarly well across all three cultural groups, which makes sense given that their age beliefs were not yet self-relevant and so would not have affected them.[16]

One of the reasons that we found such strong associations between cultural beliefs and our older participants' memory scores is that the elderly members of the mainland Chinese and American

Deaf cultures grew up at a time when they had little access to American mainstream media with its predominance of negative age beliefs: the American Deaf because they didn't yet have closed-caption TV; the mainland Chinese because of their geographic and political isolation from America; and both cultures because social media, with its ability to spread ageism across borders, had not yet been invented when the participants were growing up. Yet, within all three groups, including the hearing Americans, more-positive age beliefs predicted higher memory scores.

What our study indicated to me was that cultural beliefs about aging are strong enough to hijack memory performance in later life.

Building a Memory Cathedral

To better understand the role of age beliefs in keeping our memories sharp, I met with John Basinger, an eighty-four-year-old retired theater actor who lives half an hour away from me, in the college town of Middletown, Connecticut. His wife, Jeanine, taught film studies at Wesleyan University for six decades. She basically invented the discipline and is now an iconic figure both on campus and in Hollywood. Although John's work and legacy are a little different, his exploits loom just as large over Middletown.

Back in 1992, when John was on the cusp of turning sixty, he challenged himself to memorize John Milton's *Paradise Lost*, the epic lyric eighteenth-century poem about the temptation of Adam and Eve by Satan and their subsequent expulsion from Eden. John started slow, learning seven lines at a time, while exercising at the campus gym. He didn't think he'd memorize the whole thing. But when John starts something, he usually finishes it, no matter how long it takes. Eight years later, close to the end of his seventh decade and at the dawn of a new millennium, John had the vast, epic, sixty-thousand-word poem committed to memory. That's about the length of a full novel like *Lord of the Flies*! Then he performed the poem in an extraordinary recital that lasted all of three days.

Twenty years later, he says he still remembers all of it. The morning of our meeting, he recited one of the poem's twelve books in its entirety as a mental warm-up. But John isn't a one-trick pony. In recent years, he memorized large chunks of *King Lear*, the Shakespearean drama about an aging monarch, to turn it into a one-man show. Not long ago, he also memorized Lord Tennyson's rowdy poem "The Charge of the Light Brigade," and then set it to rock 'n' roll music and performed it with a rowdy band.

In our conversations, John insists that his memory is nothing above average. His wife and daughter have "naturally good memories," he told me, whereas he is someone who is helpless without a to-do list and often tends to forget things, like the whereabouts of his appointment book. And clinically, John is absolutely right: his memory isn't above average. His memory for everyday tasks is completely normal, according to John Seamon, a Wesleyan University psychologist who became fascinated with John's exploits and ran a battery of tests to figure out how he did it. Seamon's conclusion: "Exceptional memorizers are made, not born."[17]

John is living proof that a completely average memory is a remarkable thing when joined with the willingness to work it like a muscle and the right set of age beliefs. An image that often comes to mind, he told me, is of Pablo Casals, the great Spanish cellist who performed well into his nineties. Toward the end of his life, Casals had trouble walking and getting around, but as soon as he sat down to play, John said Casals became as fluid and graceful as he'd been as a young man.

I was curious about how exactly John had memorized a poem the size of a novel and asked about the technique he used to pull off this feat. It turned out he developed his memorization strategy almost by accident. It was during his formative years, he told me, back when he was working in theater among the Deaf.

I sat up when I heard this. "But you're not Deaf, are you?" I said, wondering if I had overlooked an obvious aspect of his identity and a thrilling connection to Deaf culture and its positive age beliefs.

John smiled and shook his head, but he began to use sign language as he told me his story. As a young man, he wanted desperately to be in theater, and the first job that came around was in sound design, with the National Theater for the Deaf (NTD), in Waterford, Connecticut, in the 1960s. The NTD was founded by David Hays, a successful set designer who had worked with Elia Kazan and George Balanchine—film and ballet greats, respectively. This new group would pioneer a novel kind of performance featuring both Deaf and hearing actors who signed and mimed and spoke their lines all at once, engaging with all the senses and making for a radical new kind of theater for both Deaf and hearing audiences. It was electrifying. John toured with the company for three years, eventually moving on from sound design to acting and teaching theater and American Sign Language.

As he told me about his beginnings, John continued signing to illustrate how performances worked at the NTD, as well as to explain what he had discovered about the process of memorization. As he worked in Deaf theater, John noticed that he was memorizing his lines more easily when he added gestures to make the spoken text more visual. When he began to memorize *Paradise Lost* decades later, he returned to that idea of making the text physical by "adding natural gestures to the mix." This, he explained, was what allowed him "to occupy at once the emotional and physical space of the poem."

John's time with the NTD, where he was exposed to Deaf culture, clearly had an impact on him. You'll remember from my study of the Deaf culture that younger members often treat older members as role models and leaders.[18] John's exposure to this culture reinforced his positive age beliefs and taught him a great deal besides. He went in knowing nothing about sign language; a few years later, he was *teaching* it. John eventually left the troupe to spend more time with his family instead of touring the country for months on end. But his experience with the Deaf stayed with him forever. His own crucial guidance, he seemed to now be suggesting, came from the Deaf, whose culture became, for a while at least, his own, and which he leans on, decades later, to pull off his remarkable memory feat.

When he talks about his life, John constantly quotes from movies, books, and poems, many of which I was embarrassed not to have seen or read, but which I later tracked down. Some of these provide him with additional positive images of aging. One novel he says he worshipped as a young man was *The Way of All Flesh*, a Victorian novel by Samuel Butler that indicted the hypocritical value system of that era. He mentioned his two favorite characters—Alethea, the loving aunt, and Overton, the novel's narrator—older figures who fit in nicely, he said, with the mythological archetype of the wise old woman or man who gives crucial guidance.

John's life is a reminder that memory is not the fixed, finite neurological resource we often take it to be. It isn't something you have or don't have, except in cases of neurological deterioration, such as Alzheimer's disease—but even then, as we will see, memory loss isn't always a foregone conclusion. Memory is malleable, and it can be enhanced. In fact, John made extraordinary use of the type of memory that most cognitive literature says should decline in later life: episodic memory.

Improving Memory by Shifting Age Beliefs in the Lab

After our cross-cultural study with Chinese and Deaf American participants, I suspected that cultural beliefs played a crucial role in memory health but knew that I would need to study age beliefs in a more controlled setting to prove just how powerful they are. So I tried to come up with a way to experimentally reproduce what I thought might be happening across these three different cultures.

After pilot testing a number of techniques to activate age stereotypes in older people, I decided to try implicit priming—a technique that had previously been used to examine racism by subliminally activating the stereotypes about Black people that are held by white college students.[19] I wanted to try something different and see if I could activate self-stereotypes, or stereotypes about one's own group. And I wanted to try this with older participants, even though a neuroscientist in my department told me that it was likely to fail since older people have slower processing speeds. I was delighted when the technique proved successful, even with one of our participants, a ninety-two-year-old man, using a computer for the first time.

The reason unconscious priming works so well is that it is able to bypass the psychological strategies we use to protect our existing positive or negative age beliefs. For example, we see this in what is called the "confirmation bias" that operates by giving more weight to evidence that confirms what we already think is true and undervalues evidence that could disprove it.

My team and I recruited a group of older participants and brought them into our lab in the Harvard psychology department. After sitting them down in front of computer screens, we told them to focus on a bull's-eye as words flashed just above or below it, fast enough for them to experience what is called "perception without awareness," but slow enough to be perceived and absorbed. What the participants thought were passing blurs were words associated with positive or negative stereotypes of old age, such as "wise," "alert," and "learned," or, conversely, "Alzheimer's," "senile," and "confused."

Before and after these priming sessions, participants performed the very same memory task used in my study with the three cultural groups, such as studying patterns of dots on a grid and then reproducing them with a stack of yellow dots on a new blank grid. I wanted to find out whether priming techniques could be used to tweak our participants' views on aging, and if these could harm or improve the types of memory that we assume decline in later life.

The result? Participants who had been primed with positive age stereotypes for just ten minutes improved their memory performance. Ten minutes of negative age priming saw a comparable decline. We found the same pattern of results in the participants, whether men or women, sixty or ninety, high school dropouts or med school graduates, whether they lived in the country or in a city, and whether it was their first time behind a computer screen or they were skilled programmers.[20] Since then our age belief–memory performance findings have been replicated by many other researchers, in places as far as Korea,[21] seven thousand miles from our American lab. Studies in five continents have confirmed the universal nature of our pattern of results.[22]

Consider the implication: although aging is a biological process, it is also a deeply social and psychological one. Your view of aging and its impacts on memory can actually influence your memory's health and performance. And those views can be shifted along negative or positive lines.

This is why someone like John Basinger, whose memory is otherwise "normal," was able to train himself to memorize a poem the size of a novel. I asked John what motivated him, all those years he devoted to building something so vast and grand inside his memory. He said it was the Greek ideal of a strong mind inside a strong body, as well as the belief that advanced age is a time when layers of knowledge and practice can yield great rewards. The poem, *Paradise Lost*, was now like "a cathedral that I carry around in my mind." He told me he often felt a little like Pablo Casals, performing all those beautiful Bach suites into his late nineties. For John, too, the music flows, and the cathedral stands inside the glorious hollows of his brain.

Back to the Future: Memory Across the Life Span

Clearly, when it comes to memory, culture matters. But our studies took place during one day. I was curious if age beliefs could impact memory across the life span. To take a closer look at this, I had to find a way to pin down people's age beliefs from decades ago, and track their memories going forward, over time. I was telling my family about this challenge one morning when my daughter suggested I use the time machine from her favorite movie, *Back to the Future*, to travel back in time and find out people's age beliefs before returning to the present, four decades later, to measure their memories.

My friend Robert Butler came up with a somewhat similar, but more feasible, solution. As the founder of the National Institute on Aging, he helped launch the Baltimore Longitudinal Study of Aging (known as the BLSA), the world's longest-running study of aging, beginning in 1958, and still in operation. Every two years, participants in the BLSA complete a bevy of tests and questionnaires to help scientists study virtually every aspect of aging they can think of. In one of these measures, the interviewer presented ten geometric-figure cards, each for ten seconds, removed each card, and then asked participants to draw each shape from memory.

Robert said he thought one of the original investigators of the BLSA had also included a view-of-aging questionnaire that he didn't think anyone had ever studied.

To find out if Robert was correct, I called the scientific director of the National Institute on Aging (who later became a valuable collaborator), Dr. Luigi Ferrucci, to pitch him my idea to connect age beliefs to memory over time and ask if he knew how I could find out if the BLSA pàrticipants ever responded to an age-belief measure. He said it was possible that one of the early investigators had included such a measure, but he wasn't sure as he didn't think anybody had published a study using it. He sent me a manual the size of a phone book of a large city to check. I was delighted to come across a measure of age beliefs that was included in the first survey of participants. It wasn't labeled, but after some more searching, I found it was called the Attitude Toward Older Persons scale.

Now, much like the character played by Michael J. Fox in *Back to the Future*, I could go back in time and examine how the BLSA participants described their views of aging at the start of the study, decades before they reached old age—many were young adults. By now, they had all reached their sixtieth birthday. I matched their age beliefs from the start of the study to their memory scores over the next thirty-eight years and discovered that people who held positive age beliefs from the outset went on to experience 30 percent better memory scores in old age than their peers with negative age beliefs. The beneficial impact of participants' positive age beliefs on their memory was even greater than the influence of other factors on memory such as age, physical health, and years of education.[23]

Now that I had studied aging and memory in three different cultures and looked at age beliefs in a lab and over time, I found evidence that aging is not the only factor that influences memory; age beliefs affect our memories to an astonishing degree.

The Gift of Age: Who Should Read an X-Ray

According to neuroscientist Daniel Levitin, certain types of memory actually *improve* with age. People are better at pattern recognition, for instance, when they get past sixty. As he puts it, "If you're going to get an x-ray, you want a 70-year-old radiologist reading it, not a 30-year-old one."[24]

As we age, our brains continue to make new connections. Angela Gutchess, a neuroscientist at Brandeis University, found that the aging brain tends to be less specialized in drawing on its many regions, which can be a good thing. Her elegant MRI brain studies showed that when memorizing a poem or other verbal information, young adults draw on the left frontal cortex. Older adults tend to use not only this same region but also the right frontal cortex, which is typically used for storing and processing spatial information, such as a map. Relying more on both brain hemispheres is a mark of adaptiveness and flexibility.[25]

Remember that John Basinger's incredible feat of memory, which would make individuals of any age extremely proud, was accomplished in part by using hand gestures that made the text more spatial—as well as by embracing aging as a period of accrued skills and experience. Age beliefs don't exist in a vacuum; they occupy the thrones of our minds, which are the control rooms for our bodies. They are part and parcel of how we code aging. They affect how we, as a culture and individuals, design, structure, and experience old age. This is why their effects ripple out in such significant ways, changing not just how we remember, but how we behave, including whether we pass on our knowledge to others.

Passing on Memories:
Mushroom Hunting in the Redwoods

Patrick Hamilton lives in the cool, lush forests of Northern California, where he has foraged, cooked, sold, studied, and taught about mushrooms for the last three decades. A sturdy seventy-three-year-old gray-haired man with a soothing voice and an easy smile, he has become a leading authority on mushroom hunting. His name comes up in numerous guides and websites devoted to mushroom foraging, and there's a picture of him proudly posing with two bowling ball–size porcinis on the back of *All That the Rain Promises and More*, the legendary mushroom-hunting guide.[26]

I sought out Patrick, though, not for tips on where exactly to find chanterelles in my Connecticut backyard, but to learn the secrets of the memory of older mushroom hunters. Patrick can identify thousands of species. This is particularly impressive considering that many mushrooms have ever-evolving characteristics. Most mushrooms grow extremely fast, changing shape and color as they age.

I also had personal reasons for wanting to know more about mushroom hunting. When I was growing up in Vancouver, our house was a few doors down from an elderly Chinese woman who lived with her grown son and his family. Every spring, on weekend mornings, they would all head off into the woods with baskets. If I was in the yard when they returned in the afternoons, the jubilant family would invite me over to show off the piles of fantastically colored mushrooms that the matriarch had unearthed. Grandma Leung seemed to have a magical sense for where to find mushrooms that her son and his family admired to no end.

Patrick's vast memory for mushrooms has a debt to positive age beliefs. Like the actor John, he demonstrates that these beliefs can contribute to episodic memory. This is a type of detailed memory of events and objects that many scientists wrongly assume always declines with age. Patrick explained to me that he has observed, "As I get older, I'm gaining wisdom."

As a child, Patrick fell in love with the outdoors while playing with his grandparents in the sweet-smelling orange grove they bought near Los Angeles, after emigrating from Ireland. At the age of forty-three, he "fell into mushrooms," when he moved to a house on the Russian River, the redwood-lined estuary that runs through Mendocino and Sonoma Counties and drains into the Pacific. At the time, Patrick knew two edible mushrooms by sight—coral mushrooms and chanterelles. He and his wife would go out into the forest with baskets and knives and come back with dinner. Since then he's lived all over Northern California—from Point Reyes on the coast to the moist redwoods of the Sierra Nevada—working as a chef, teacher, supplier for fine San Francisco restaurants, and foraging columnist for a local newspaper.

When I asked him what pictures come to mind when he envisions aging, he prefaced his response by saying that he doesn't picture older people in the city, but rather outdoors, in the woods. He then suggested two images. One was of an attractive woman in her seventies with long gray hair enjoying herself on a long hike. The second was of an older man on a mountain bike, gleefully rattling down a narrow path between towering trees.

Patrick draws on these images of older persons enjoying and immersed in nature in his own life. At seventy-three, he is older than most people who join his classes in the woods. But he's also spryer than most. Recently, he and his foraging partner, a seventy-nine-year-old fellow mushroom enthusiast, went out for a hike and got caught in heavy snow. They had to hike seven miles through the Sierra without cell-phone coverage. They couldn't stop, he told me, or they'd risk becoming frostbitten. They made it home in one piece, and despite the skin-tingling challenges of the experience, Patrick made the episode sound like an adventure he would gladly repeat, if given the opportunity.

Mushroom hunting also reinforces positive age beliefs in important ways, because elders tend to have more of this specialized knowledge and are looked up to by younger foragers in what is frequently an intergenerational activity. It's also an activity where accumulated knowledge—and who holds on to it—can be a matter of life and death, since many mushrooms are poisonous.

"My age actually helps me when it comes to finding mushrooms," Patrick explained. His ability to identify thousands of different species is based on his many years of hunting for them in forests and deserts as well as spending a lot of time reading scientific texts. "Learning to identify different mushrooms isn't just a question of committing the name and image, not to mention the toxicity or edibility, to memory, since the same mushroom looks so different at various stages of its life, and in different climates. It's also about understanding and learning the subtle interactions and interdependencies of the broader context. I look at the soil, the trees, the plants, the geology, the whole thing." He seems to apply these beliefs to people, too, and to the way he lives his life: nothing exists independently of anything else; certain cycles are dependent on other cycles; living things never stop growing.

Patrick's students tend to be couples, or parents with their kids. Sometimes the grandparents join his trips into the woods. There's something special when the whole family is there: "Everyone's on the same level in the forest, and you don't think about anyone's age." He'll often find a giant, hollowed-out redwood stump for the whole family to pose on for pictures.

Recently, he told me, he was out foraging with a Russian family when he noticed the grandmother furiously collecting bright red russula mushrooms. She had even enlisted her grandson in the task, but Patrick was alarmed. Patrick ran over and told her this was the

Russula cremoricolor, also known as *emetica*, or "the sickener." She laughed and told him that it can be made edible by pickling (but this is not recommended for the inexperienced). Russia, it turns out, has mushrooms that are similar to those of Northern California's forests, and the Russians can be fearless cooks. By the end of the exchange, the woman's grandson was beaming at her. Patrick gave the grandmother his blessings. Then she and her grandson returned to the forest floor.

Elders' Songlines as a Path to Cultural Survival

Patrick's role as a keeper of knowledge reminded me of the anthropologist Margaret Mead's writings about the role of the older members of Indigenous cultures around the world.[27] She observed that they were often indispensable due to their immense recall of cultural knowledge, which they shared with younger generations. This memory-keeping role, in turn, helped solidify their social standing: "Their tireless industry represented physical as well as cultural survival. For such cultures to be perpetuated, the old were needed, not only to guide the group to seldom-sought refuge in time of famine, but also to provide the complete model of what life was."[28]

As an example, Mead mentions the Indigenous Australian culture, which thrived in an extremely harsh landscape for thousands of years. These Indigenous people have survived to this day because of their traditions, which unite them with the local ecology. The elders preserve and pass down intimate and indispensable knowledge through Songlines—vehicles of cultural and informational transmission that include a corpus of songs that provides the oldest continuous oral history of any group of people on Earth.[29] These songs function as cultural and spiritual "texts," as well as oral encyclopedias that contain information about the thousands of species of plants and animals across Australia, and detailed maps of hundreds of kilometers of land.[30]

Mead writes, "The continuity of all cultures depends on the living presence of at least three generations."[31] This observation applies to the elders who teach the Songlines that help their children and grandchildren find shelter and food, and to Patrick, who shares with mushroom hunters of all ages his gastronomic and often life-saving knowledge about which mushrooms are poisonous. These instances of elders being valued for transferring information to younger generations are true "senior moments": in which "senior" represents remembering.

3

Old and Fast

In a hilly green corner of the northwestern United States lives a cheerful nun who has become a bit of a local celebrity in the Spokane Valley. That's because since 1982 Sister Madonna Buder has completed over 350 triathlons, earning the nickname "The Iron Nun." The first time she went on a run, she was nearly fifty years old, using a pair of sneakers borrowed from a friend. Now, she's ninety-one and has just completed another triathlon. Her training is straightforward and unconventional. No fancy gyms or Olympic coaches—instead, Sister Madonna runs or bikes to the grocery store, swims in the local YMCA pool, and uses snowshoes to get around in winter. And for all her physical prowess, Sister Madonna looks rather unassuming, with a slight build, bright blue eyes, and a short haircut that she does herself.

We were talking about aging when I asked her what the first five words were that came to mind when she thought of "an old person." She responded, "Wisdom and grace," and thought some more. "There's running," she said, "and opportunity." When I pointed out that was only four, she laughed, and added, "Fine wine."

She credits her grandmother for helping her associate aging with opportunity and wisdom. Back in Saint Louis, when she was a young girl pondering the vastness of life and wondering what to do when she grew up, her grandmother told her: "The past is dead and gone, the future stops being the future once it gets here, so the present moment's the only thing you're responsible for." Eighty years later, it's how she thinks about aging: "It doesn't make sense to fear aging, since you never know what lies ahead because you have never had this experience before."

Her father was a champion oarsman who rowed and played handball into his seventies. She remembers being proud of him for being so athletic at what she considered, back then, to be an advanced old age. When she herself entered her seventies she competed in the Ironman World Championship on Hawaii's Big Island. After she biked across the island and swam nearly three miles of open ocean, she was running the marathon component when she started thinking of her father. It was a full moon; the road was covered with silver-tipped shadows, and she remembers feeling his presence as though he were running alongside her.

Functional Health Discoveries

When I met Sister Madonna, I marveled at how effortlessly she disproved the stereotype of debility and decline so often considered to be the natural trajectory of aging. I also wondered whether her positive age beliefs could be driving her physical accomplishments. To find out whether age beliefs can influence functional health, which is how well our bodies move based on things like gait, balance, stamina, and speed, I designed a study. I examined an existing survey that asked its participants, all aged fifty and older, their age beliefs—for instance, whether they agreed with statements such as, "As you get older, you are less useful." Their answers were given scores indicating negative or positive age beliefs. Over the next two decades, these participants had been tested on their functional health every few years.

I found that the participants with positive age beliefs showed much better functional health over an eighteen-year period than those of the same age with negative age beliefs.[1] This was the first time that anybody had demonstrated that age beliefs—rather than "aging"—were a major factor in later-life physical performance.

But I still had to be sure that the cause and effect were not the other way around: that it wasn't better functional health that led to positive age beliefs. I talked to a statistician and good friend, Marty Slade, about this very question. Marty was an aeronautical engineer before he became a statistician. He uses the same logic and brilliance in his analyses that he did to test airplane engines. We first checked whether a reverse association existed. That is, whether functional health when participants joined the study predicted age beliefs over time. It did not. Then we looked at all the participants who had the same functional health score when they joined the study and repeated the same analysis. We found that age beliefs predicted functional health, rather than the other way around. More recently, research studies in other countries, including Australia, have found the same results.[2]

Physical Prowess as a Snowball Effect

To make sure that age beliefs were behind better physical function, my team invited older participants to come into our lab. We randomly assigned them to either a positive or negative age-belief group, and primed them, as we had in the earlier memory experiment, with subliminal positive or negative age beliefs. Participants who were primed with positive beliefs for just ten minutes immediately showed faster walking speeds and better balance, which was measured by greater time with their feet lifted off the ground with each step (calibrated by special pads in their shoes to record pressure). Those who were exposed to positive beliefs of aging walked *better*.[3]

As my goal has been to harness positive age beliefs to bring about health improvements of older persons in the community, the next step was to see if we could bring about the same improvement outside our lab: in different senior housing complexes. Further, in our lab study, although we found that priming impacted participants' walking right away, the effect started to fade after an hour. I followed up with another study that primed elderly participants, at one-week intervals for one month, hoping it would yield more lasting improvement.[4] This new study got closer to how age stereotypes operate in the real world, where we're regularly exposed to stereotypes over long periods of time.

An eighty-three-year-old woman named Barbara was one of our participants. She had seen a flier for the study in the community room of the complex where she lived. The study requirements included that she respond to a few surveys, engage in some computer games, and do a few physical exercises. Well, she thought, why not? It was something new to try out.

During her month of weekly sessions, Barbara met with one of our research nurses and sat in front of a computer screen to do a few simple speed-reflex games before moving on to things like sitting down and getting up from a chair five times in a row, walking across a room and then back, and standing with one foot behind the other for ten seconds.

Initially Barbara had difficulties. When getting out of the chair five times in a row, she felt like she was about to fall and might need to grab the hand of the research nurse nearby. And balancing with one foot behind the other for ten seconds, as if on a tightrope, is not as simple as it sounds.

In the third week, however, something interesting happened. Barbara felt more confident getting in and out of that chair. And she said that standing still with her feet one behind the other no longer made her feel like the Leaning Tower of Pisa.

There were other changes, too, subtle yet noticeable. She reported that it seemed easier to get out of bed in the morning and easier to climb the front steps of the public library where she likes to rent DVDs. And it wasn't just these changes—Barbara felt better overall, more in charge.

She surprised herself, for instance, by calling a cousin she hadn't spoken to in years. On the spur of the moment, she signed up to organize the entertainment for the upcoming holiday party her senior housing complex was hosting. She joined a playwriting group in her building that performed short pieces at a nearby theater.

If you are wondering if the reason that Barbara's balance and mood improved has something to do with our intervention, which strengthened her positive age beliefs, you would be right. By subliminally prompting her with words like "spry" and "fit," we activated her deeply ingrained positive views about what it's like to be old—and moved them to the forefront of her belief system, which for so long had been dominated by negative images of aging assimilated from society. This activation then improved Barbara's perception of herself as an older person, as well as her physical functioning.

Of course, Barbara didn't grow wings overnight. But her physical improvement after a month of priming was similar to what someone the same age tends to register after six months of exercising four times a week.[5] And she was far from the exception: after a month, the group of older participants exposed to positive age beliefs walked significantly faster and had better balance than the neutral-comparison group of older participants who were exposed to random letters.

Unexpectedly, my team found a snowball effect. That is, like a snowball growing while being rolled down a snowy hill, the beneficial influence of positive age beliefs on our participants' physical function steadily grew over the two months of the study.[6] A virtuous cycle was at work: positive priming strengthened participants' positive beliefs about older people, which strengthened their positive beliefs about their own aging, which strengthened their physical function, which further strengthened their positive age beliefs. This, in turn, led to further strengthening of their physical function. (See Figure 1 below.)

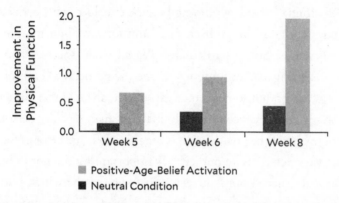

Figure 1: Activating Positive Age Beliefs Improves Older Persons' Physical Function over Time. Those participants exposed to positive age beliefs showed significantly better physical function than those in the neutral group; this beneficial influence of positive age beliefs grew over the two months of the study.

Swimming Ahead from Within a Group

Age beliefs exist along a continuum, but most people in the US are mainly exposed to and express negative ones.[7] Other people, like Sister Madonna, seem to draw mostly on positive beliefs. But having positive age beliefs doesn't mean you have to be extraordinary, competing in triathlons every month. Sometimes they can just help you try something new.

Take Wilhelmina Delco, a Texas politician who picked up swimming for the first time at eighty, ten years ago, and who is often the oldest person in the pool. People in the neighborhood recognize her as the "old lady who swims at the Y," she says with a laugh, but her achievements are much more significant than that. Three days after the assassination of Martin Luther King Jr., she became the first Black person elected to public office in Austin when she joined the Independent School District Board of Trustees. Since then she has given four decades of trailblazing civic service as a state representative in the Texas legislature.

Wilhelmina started swimming because it helped with her arthritis, and it felt good "to go backwards and forwards" during the ten laps she swims. She is particularly pleased when she sees another octo- or septuagenarian jumping off the diving board. "I am proud of being my age, not some strange exception," she told me. Similarly, even though she has been the only Black woman or older person or both on many of the boards of civic organizations she belongs to, she doesn't view herself as an outlier. "It's important for me not to be an exceptional other person." It seems that she sees herself as leading from within the group, rather than way ahead of others. Instead of preaching solidarity, she channels it through her example.

This ability to show others that if she can do it, they can, too, comes in part from her experiences dealing with racism. For most of her life, she has felt unwelcome in many of the places where she's worked and lived. And even though being targeted by prejudice on multiple fronts—being Black, being a woman, being older—can cause an exponential increase in stress, it can also offer opportunities to transfer coping strategies across identities.

Wilhelmina's ability to navigate racism has helped her navigate ageism. Her mother, who lived with Wilhelmina's family for the last twenty years of her life, taught Wilhelmina a great deal about how to thrive in a racist environment. If something needed to be done, her mother would say, "You should do it. Bigots be damned. You ought to try."

Wilhelmina also learned about aging from her mother—and from her mother-in-law, who also moved in with the Delcos in her later years. Having two grandmothers in the house was a great source of support and love for Wilhelmina and her children ("And quibbling," she added with a chuckle) and impacted the way she thinks about old age. They often gave her and her children advice on everything from cooking to how to balance a checkbook to confronting challenges, and finding strength in family and community ties. Now that Wilhelmina is the age that the matriarchs in her life once were, she insists she'd never lie about her age: "I'm too glad to have made it."

Swimming to Late-Life Gold

After meeting Wilhelmina, I spoke to another woman who discovered the pleasures of swimming late in life, although once she got in the pool, this one basically never got out. Nearly every passing year, it seems ninety-nine-year-old Maurine Kornfeld breaks another world swimming record.

It started when she was ninety, and since then she has set twenty-seven such records.[8] When she was in her sixties and a social worker in Los Angeles, the local YMCA pool that she used to cool off in the summertime closed, so she went to the next closest pool. Only Saturday mornings were available because a team of Masters swimmers that belonged to a national swimming club used it all the other mornings. When she called the coach to complain about his monopoly over the pool, he asked her questions she didn't understand, like "What's your stroke?" "I had no idea what he was asking," she laughs, "but he said to come on Saturday morning and he would look me over, so I went that Saturday morning and he said get in and swim freestyle, and I got in and swam with my head up, because that was sensible. He kept shouting at me to put my head in the water, which seemed like a vulgar idea. He'd been a sergeant in the marines and kept shouting at me, but I swam backstroke after that and he liked what he saw and, like that, I was the latest recruit, and I'd just landed on Parris Island." She smiled, referring to the notorious base where the US Marines train their recruits.

Soon afterward, Maurine started competing in Masters swim meets. And soon, she was winning all of them. But the joy of it, Maurine insists, derives from swimming, rather than the competitive element. "What I love are the people, the pleasure. Swimming makes you feel good, it soups up the endorphins. It's such a lovely, sensuous sport. It doesn't really matter whether you compete. When you're in the water, you feel immortal. Nothing can touch you. You just feel good."

And Maurine is someone who makes others, including me, feel good. Her presence is curious, lively, and joyful. She constantly turns the conversation back to others, more interested in hearing about their lives than in talking at length about herself. After our first meeting, she immediately began sending me weekly emails with updates on her life, helpful suggestions about people I might want to talk to, and videos (such as the latest in *Puppies Licking People*) that I couldn't resist sending to others to also make them laugh.

It was the middle of the COVID-19 pandemic when we spoke, and shelter-in-place orders were in effect across California. Yet the hardest part of it, for Maurine, wasn't being confined to land or missing her daily routine of getting up at 5 a.m. to drive across town in the dark to the Rose Bowl pool; it was not being able to see her friends. Old age has been a rewarding and an extremely social chapter of her life. She's been busier than when she was working, she told me, as one of the first social workers to be licensed in California. She is a docent at several Los Angeles museums and historic sites, and, though she lives alone, she sees friends constantly. When we later spoke, one of them stopped by to borrow a catalog for an art exhibit.

Maurine reads avidly on her front porch. As she explains, "I was born with a library card in my mouth—there were no silver spoons where I lived." In fact, we had to end a conversation a little early so she could finish reading the new Erik Larson book, since she was leading the next book club discussion. To keep in shape when she couldn't swim during the pandemic, Maurine walked around nearby Bronson Canyon, marched up and down the steps of her porch eleven times in a row, and lifted cans of stewed tomatoes to work her arm muscles.

After our first meeting, Maurine emailed me a line from a Robert Browning poem that captured her thoughts on later life: "Grow old along with me! / The best is yet to be."

Moving the Starting Line

What do these two very different swimmers living on opposite ends of the US tell us? Together, they show that it is never too late to start exercising, that the aging body responds extraordinarily well to exercise, and that positive age beliefs have numerous trickle-down effects, including better functional health.

It turns out that age beliefs might even be a better determinant of later functional health than whether you exercised in your youth. Jessica Piasecki, a thirty-year-old British researcher at the University of Nottingham, was one of the investigators of a recent study that found people who started running in their fifties can be just as fit and healthy as competitive older runners who have been doing it for many decades.[9] People who started running thirty years later than lifelong athletes had very similar finishing times, muscle mass, and body fat.

While Jessica's study hadn't directly measured age beliefs, since her team focused on physiological predictors, the findings changed her attitude toward aging, as well as her own running regime.

Jessica is an endurance athlete who is humble, in addition to talented. After our conversation, I found out that since she started studying older athletes, she has become the fastest British woman runner currently competing, and the third-fastest woman in British marathon history. Her respect for people who exercise later in life has only grown since her research began. When she talks to Masters runners, they all seem to have a view of aging that includes pushing themselves. And working with them, she tells me, gives her a great deal of extra motivation in her own running.

Although I am certainly not the fastest swimmer or runner (when I enter races, my goal is to finish), I can relate to feeling inspired by older athletes. One of my favorite things about being a gerontologist, or someone who studies aging, is that I often get to meet inspiring older people. The other morning, when I was trying to decide whether to hit snooze and pull the covers over my head for a bit more sleep or to get out of bed and run around the cornfields near my house before work, the exuberant, determined voices of Wilhelmina, Barbara, and Maurine popped into my head, and I rolled out of bed to go find my sneakers.

From Car Crash to Fighting Structural Ageism

Positive age beliefs not only offer the possibility of greater functional health in older people, they help them recover from illnesses and injuries, too. There's no question that it is part of life to occasionally get sick or injured. What is up for debate is why people with the same injury show different patterns of recovery.

Just as there is a widespread false belief that functional health inevitably declines with age, it is also assumed that older people do not recover well in the wake of acute injuries or illness. In an important study that upended this assumption, geriatrician Tom Gill discovered that the reason for this false belief had to do with flawed methodology: most of the studies dealing with older people and disability tracked participants every year, or every few years, which could miss brief episodes of disability and recovery, say a sprained ankle that gets better within a month. Using the longer timescale of surveys conducted a year or more apart, most researchers had charted a worsening of health conditions with little recovery. But when Gill surveyed participants in shorter intervals of every month, he found that 81 percent showed a complete recovery within one year of their initial disability episode, and 57 percent of those recovered had maintained their independence for at least six months afterward. That is, most older people who couldn't bathe or feed themselves after a bad fall or injury were eventually able to do these things again.[10]

We know thanks to Gill and his team that most older people who have even acute injuries or accidents completely recover. But what drives these recoveries?

I wondered if it could be age beliefs. Fortunately, when Gill was planning his study of New Haven residents aged seventy or older, I asked him if I could examine his participants' age stereotypes using my Image of Aging measure. To do this, at the start of the study, his team asked 598 participants to name the first five words that came to mind when they thought about an older person. We then checked in with the participants every month over the next ten years to find out if they had experienced any new injury or illness, and, if so, whether they'd partially or completely recovered.

What we found was that those who started with positive age beliefs were significantly likelier to recover from injuries over the next ten years. This age-belief pattern existed above and beyond the influence of age, sex, race, education, chronic illness, depressive symptoms, and physical frailty on recovery. Although I predicted that positive age beliefs could act as a resource for recovery, I was surprised at the magnitude of the effect: our participants holding positive age stereotypes had a 44 percent greater likelihood of complete recovery from severe disability than those who clung to negative age stereotypes.[11]

Like a ship's mast in a storm, age beliefs can be a source of security and strength as older people go through disability and eventual recovery. Take the case of Oscar-winning actor Morgan Freeman. He was driving his Nissan Maxima along a Mississippi highway one hot summer night when he suddenly lost control of the car, which flipped over several times, instantly turning into a twisted metal heap. Unfortunately, the airbags didn't deploy. The jaws of life had to be used to extract Freeman, whose crushed body was airlifted to the nearest hospital with numerous broken bones. His fans and loved ones prayed for his recovery, assuming that if he survived, the seventy-one-year-old actor would be permanently paralyzed.[12]

But Morgan Freeman not only recovered—he thrived. Since his recovery he has starred in thirty-seven more TV shows and movies. He takes special pride in starring in action movies, such as *Red* and *Going in Style*, that feature older characters as heroes kicking ass and taking names. In the latter movie he joins Alan Arkin and Michael Caine playing retirees who, after having their pensions canceled by the company where they all worked, fight structural ageism by robbing the same bank that sets up fraudulent deals with older customers. (I will discuss legal means to fight ageism later in the book.)

Today Freeman is eighty-four, enjoying old age, exploring his spirituality (he recently produced and narrated a documentary about world religions), and doing what he loves—making movies. He explained, "I can afford to retire, but now I work just for fun."[13]

In an interview that took place eight years after his car crash, he was asked, "Now as a Hollywood legend and a leading man, do you feel your work is challenging stereotypes about aging at all?" Freeman responded, "I hope so. I really hope so."[14]

Morgan Freeman, who associates old age with curiosity and vitality and who has demonstrated physical resilience in his later life, exemplifies what we found in our New Haven study. But you don't have to be a movie star, triathlete, or world-record-holding swimmer to age healthily. Whether you decide to start going for runs at sixty, hop in the pool for the first time at seventy, or go on walks at any age, it matters less when and what you do than that you build up positive age beliefs and trust that your body will respond in kind.

4

Brawny Brains:
Genes Aren't Destiny

One fall day, a college biology professor called my grandfather, then a lowly freshman, into his office and demanded to know how he'd done so well on the final exam.

"Levy," he said, holding up my grandfather's exam like it was a particularly incriminating piece of evidence, "this exam is perfect. No one ever hands in a perfect exam." The professor, whose class was a notoriously challenging one that explored cutting-edge advancements in biology, sat stunned as my grandfather began to recite sentence after sentence of the relevant textbook sections. By the time my grandfather finished, he was smiling with relief. Now, the professor understood why this student had received a perfect score. For the rest of his life, my grandfather's photographic memory would continue to impress everyone who witnessed it, including, of course, his grandchildren.

The child of poor Lithuanian immigrants, my grandfather was endowed with a great deal of good fortune. He was the first in his family to go to college. Then he attended law school. When I was young, he read me Horatio Alger novels from his childhood about youths living in poverty who rose up the ranks of society through "luck and pluck." He knew he, too, was lucky, so he made a career that aimed to bring joy to others. He started a publishing company that churned out the kind of brightly colored comic books that kids were gleefully hooked on in those days, full of oozing monsters and superheroes. Toward the end of his life, however, his luck ran out. One day, he and I were at lunch, and instead of reciting from memory the full menu backward, he urged me to pay attention to the tiny green people moving around beneath our table, lifting weights, straining, grunting, right there at our feet, like cartoon creatures out of his comic books.

Around that time, he began to lose his famous memory, and soon, Grandpa Ed was diagnosed with Alzheimer's disease. As his granddaughter, I was terrified by the slow progression of his illness, its ability to erase his ever-unfolding present and confine him within the frozen past.

It wasn't until I became a psychologist and started studying the topic that I began to think of the aging brain from a more removed, but also more hopeful, perspective. For most of us, our brains show certain advantages as we age and these advantages can be diminished or enhanced by the cultural factors around us.

Brain Biomarkers and a Village Without Senility

In 1901, in Frankfurt, Germany, a fifty-one-year-old woman named Auguste Deter came under the care of a doctor named Alois Alzheimer. Frau Deter had become paranoid and started hiding objects around her house. Increasingly, she seemed to be losing her memory. Her devastated and confounded husband sent her to live in an asylum, where she became a patient of Dr. Alzheimer. He was intrigued by her sad and tragic transformation. When asked to perform simple tasks, such as writing her own name, Frau Deter was unable to. "I have lost myself," she would repeat on end, to anyone who would listen and, sometimes, to no one at all.[1]

After her death, five years later, Dr. Alzheimer autopsied her brain, which had dramatically atrophied. After staining thin slices of brain tissue with silver salts, he discovered that it was riddled with abnormal deposits: amyloid plaques, which are protein clusters that build up between brain cells, and neurofibrillary tangles, which are twisted strands of protein that build up inside the brain's cells.[2] He had discovered the disease that bears his name and spent the remaining years of his life publishing articles on the subject, which were, to his dismay, largely ignored by the medical community.

For the next seventy-five years, little additional research was conducted on Alzheimer's disease, in part thanks to the erroneous assumption by doctors of the time that it was due to the inevitable hardening of the arteries that comes with age, as well as a general exclusion of older people from the burgeoning field of brain research.[3] But the disease was a ticking time bomb; today, nearly six million Americans have Alzheimer's disease, which is roughly 10 percent of the US population that is sixty-five and older.

Alzheimer's does not affect every culture equally, however. For instance, dementia was found to be five times more common in the US than in India.[4] Although the scientists who documented this speculated that the cultural differences might be due to diet, it seemed to me that age beliefs might play a role in this stark discrepancy. In India, older individuals are treated with great respect and routinely sought out for counsel on everything from financial investments to family conflicts.[5] That's a very different culture of age beliefs than the prevailing one in the US that often denigrates older persons.

A former graduate school classmate of mine, Lawrence Cohen, who now directs the University of California, Berkeley's medical anthropology program, tells the story of attending a global conference in Zagreb where an anthropologist from India gave a presentation on the longevity of the elders of a certain tribe in northeastern

India.[6] After he was done speaking, an American gerontologist asked about the prevalence of dementia among those elders. But the speaker didn't seem to understand the question. Other North American gerontologists jumped in to help. It appeared to be a problem of translation: "Senile dementia?" they suggested. "Alzheimer's disease?" But the speaker was unfamiliar with their terms. "What we mean," another American audience member tried, "is senility." Finally, the Indian anthropologist nodded to show he now understood the question. The audience relaxed; the linguistic gap was mended.

"There is no senility in this tribe," the Indian anthropologist explained. To him, this was obvious. He'd just been describing an isolated society in which the traditional Indian multigenerational family was still intact; in such a society, free of ageism, older people were well cared for, valued, and integrated into the social life of the community. Why, then, he wondered, should they become senile?

To investigate the impact of age beliefs on our brains' susceptibility to dementia, I turned again to the Baltimore Longitudinal Study of Aging, which has long been testing a group of volunteers with yearly brain scans. Members of another group volunteered to donate their brains to science, to be dissected and studied after death. All these volunteers described their age beliefs early in the study, while they were physically healthy and dementia-free, decades before their brains were scanned or dissected. What my team found was that people with negative age beliefs were much more likely to develop the telltale plaques and tangles than those with positive age beliefs.[7] In fact, their hippocampi, the part of the brain responsible for memory, shrank three times as fast.

Here we had it: age beliefs—an individual but also cultural factor—could impact the likelihood of developing these Alzheimer's biomarkers.

Overcoming Risky Genes with Cultural Beliefs

Alzheimer's, a neurodegenerative disorder that progressively kills off brain cells, has a genetic basis. That is to say, people born with a certain gene called *APOE* ɛ4 are more prone to developing Alzheimer's than others. When looking at health, genes are important. You may have heard some people say that "genes are destiny." Everything about you, this vein of thought contends, is determined by your genes. In high school biology classes, students learn about Gregor Mendel, the nineteenth-century Augustinian abbot who discovered the laws of genetic inheritance by studying and crossbreeding different varieties of pea plants. It was their genes, he found, that determined characteristics such as height or leaf color. For a long time, we thought that the same applied to humans: that genes controlled our intelligence, attractiveness, personality, and health.

Even though many of Mendel's pea-based observations became the basis of modern genetics, in the last few decades great advances have been made in a field called epigenetics. This field shows how environmental factors influence how genes determine outcomes. For instance, Gregor Mendel would have been dealing with epigenetics if he'd tried singing to half of his seeds and found that the pea plants that emerged from these musical seeds were taller than the plants grown in total silence. (As far as I know Mendel didn't try this musical experiment.)

One interesting epigenetics study shows that baby mice who are groomed, licked, and nursed more by their mothers develop new resilient genes that they pass on to their own offspring.[8] Lots of different factors can affect gene expression. Increasingly, scientists are discovering that cultural and environmental factors play an important role in determining our health. As an example, consider the asthma risk of Latino children in the United States: some of it is coded in an ancestral genetic component, but there are also environmental factors such as air pollution, which tends to be higher in minority communities, that can exacerbate genetic expression of that risk.[9]

Likewise, when it comes to Alzheimer's, I found age beliefs (among other environmental factors) can help determine how genes related to this disease are expressed.

Just as we are all born with eyes that can be brown, blue, hazel, green, or gray, we are all born with slightly different types of the *APOE* gene: the ε3, ε2, or ε4 variant. Most of us are born with the ε3 variant, which doesn't influence our susceptibility to Alzheimer's. Ten percent of us are lucky enough to be born with the ε2 variant, which protects against dementia and promotes longevity. It's the unfortunate ε4 variant that plays a role in Alzheimer's disease. About 15 percent of the population is born with that variant. What's fascinating is that only *half* of those people go on to develop Alzheimer's. Why is that?

To find an answer, I tracked a national sample of more than five thousand older people over a period of four years, and uncovered an effect that was much bigger than I expected: among participants who carried the risky *APOE* ε4 gene, those with positive age beliefs were 47 percent less likely to develop dementia than those with negative age beliefs. In fact, as you can see in Figure 2 on the next page, they had roughly the same likelihood of developing dementia as those *without* the risky gene who had positive age beliefs. In other words, biology targeted them for dementia, yet half of them never developed the disease—thanks in part to the shield provided by their positive age beliefs.[10]

This study was the first to examine whether a societal factor (age beliefs, in this case) could reduce the risk for dementia in *APOE* ε4

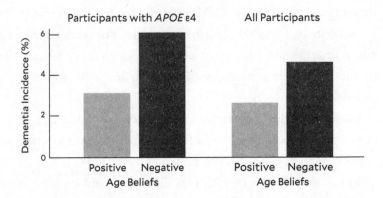

Figure 2: Positive Age Beliefs Reduce Risk of Dementia. These beliefs reduced the risk of dementia for all participants, including those with the risky gene *APOE* Ɛ4.

carriers, as well as in older individuals in general. We found the advantage of age beliefs in reducing dementia risk to be much greater than the factors most often studied as contributing to dementia risk, such as age, sex, depression, and earlier cognitive scores.

Preventing Alzheimer's Symptoms with Age Beliefs and Jamaican Seasoning

Although a vast number of people who get Alzheimer's disease develop the same set of symptoms that robbed my grandfather of his memory and personality toward the end of his life, others with the same telltale buildup of amyloid plaques in their brains see their cognition remain relatively intact. Their brains indicate the signature biomarkers of Alzheimer's, yet they express few, if any, clinical signs of the disease.

To better understand how this can happen, I spoke to several people enrolled in a major ongoing study, the Anti-Amyloid Treatment in Asymptomatic Alzheimer's Disease (A4) study, supported by the

National Institute on Aging and currently taking place in sixty sites around the US, Canada, and Australia.[11] Its goal is to come up with ways to prevent Alzheimer's symptoms before they appear. To do so, investigators have been looking at people who have one of the neurological hallmarks of Alzheimer's—elevated amyloid plaque levels—but normal cognition and no observable dementia.

I first met Amy, a retired eighty-two-year-old bookkeeper born in Jamaica and now living in Chicago, who initially got involved in the A4 study because her sister wanted to join the study but failed the screening when she was found to have symptoms of dementia. Amy volunteered to be her replacement, and although she has the telltale amyloid plaque buildup in her brain, she doesn't show any symptoms.

Six years of MRIs, questionnaires, and memory games and one uncomfortable spinal tap later, she's grateful to have a firm grasp on her own mind. These days, she leads a quiet, contented life: church, frequent telephone calls with her daughter, lots of Jamaican cooking. She grew up in the lush hills outside Montego Bay, without electricity and running water. Her father was a headmaster and church deacon, an older man in a community that treated him with attentiveness and respect. She remembers that when he was called to deal with community problems, such as falling high school graduation numbers, he frequently worked with the other village elders to come up with creative solutions.

After nearly a lifetime in the United States, Amy, who mainly socializes with other Caribbean immigrants, is routinely disturbed by the way elderly people are treated in the US. She's spent the last decade volunteering for an organization that helps kids in impoverished Chicago neighborhoods learn to read and is often shocked by how some of these children speak to the older volunteers in rude and patronizing ways. Instead of reprimanding the children, the teachers sometimes laugh at their comments.

This way of treating older persons stands in stark contrast to what Amy experienced in Jamaica. Many Caribbean cultures place respect of older persons at the top of their value systems, which often makes taking care of them an esteem-generating activity. Since she's cognitively better-off and more mobile than her older sister, Amy looks after her. Growing up, she tells me, the two of them weren't very close; but now they're best friends who appreciate each other with a depth of feeling that wasn't there before. Family fills the space that retirement has opened up in Amy's life. She sees her sister and eldest daughter as often as she can.

What is it about Amy that makes her resistant to clinical dementia, even though she has an elevated biomarker of Alzheimer's? There is reason to believe that age beliefs play a key role. Short of a cure, the most effective way to control Alzheimer's is to tamp down on stress.[12] Stress boosts inflammation in the brain, and chronic inflammation plays a role in numerous illnesses,[13] thereby paving the way for the development of neurological disease. The disease itself then disrupts the neural and endocrine pathways involved in the stress response, thus accelerating its own progression. It's a vicious cycle that is best countered by good stress management, which is why doctors are often on the lookout for stress and ways to lower it with regular exercise and healthy eating habits.

Another stress-lowering factor we should consider is age beliefs. When I was a postdoctoral fellow at Harvard Medical School, in an experimental study conducted with older Bostonians, I found that negative age stereotypes magnified stress, whereas positive age beliefs acted as a buffer against it.[14]

Because of their stress-protective properties, positive age beliefs are even helping people who carry the risky *APOE* ε4 gene to ward off the fate biology seemed to have decreed for them. It's likely what happened to Amy, who has the brain pathology of Alzheimer's, without the cognitive symptoms. Thanks in part to her positive age beliefs, she is better able to handle stress and she pushes herself to stay physically and mentally active by walking around town and doing jigsaw puzzles and crosswords. She's so good at Scrabble that anyone who plays her more than a few times refuses to make it a regular habit. Her sister occasionally tries to sneak in words in Jamaican patois, but Amy is a stickler: if it's not in the Scrabble dictionary, it's against the rules.

A healthy lifestyle doesn't have to be difficult or expensive. In Amy's case, the simple pleasures are the healthiest ones. Before the COVID-19 pandemic, she helped out however she could at her church, doing floral arrangements and bookkeeping. When we spoke, everyone was shut indoors due to the pandemic, so she spent her time cooking and bringing food to her sister. That afternoon, she was cooking flour dumplings and grilled red snapper with Jamaican jerk seasoning.

Amy's lifestyle is steeped in her positive age beliefs, which are rooted in her Jamaican upbringing. For instance, she believes to her core that old people have valuable opinions. As a result, Amy has become much more outspoken as she has aged. She was always a reserved person, hesitant to chime in during group conversations, "by no means a good talker." But as she's gotten older, she's become more outgoing, eager to give a piece of her mind. As if channeling the outspoken elders of her Jamaican childhood, these days she lets her friends and people around her know how she feels, especially if she encounters ageism. Getting to know Amy, I was reminded of something Maggie Kuhn, who founded the antiageism activist group the Gray Panthers, said: "Old age is an excellent time for outrage."[15]

Two Young Doctors Equal One Older Jonas

Jonas, a seventy-five-year-old pediatrician from the Midwest, demonstrates that old age is also an excellent time for growth. He saw a flier one day inviting people to be screened for an Alzheimer's study. At the time, he was mourning the death of his father, who had developed Alzheimer's, so he decided to give it a go. He's now a participant in the A4 study, which means that, like Amy, his Alzheimer's-disease biomarker, amyloid levels, are high, but his cognition is unaffected. Jonas is one of those people who were targeted by the disease but have managed to resist its symptoms.

A few years ago, Jonas retired from clinical practice, although he continues to teach.

"I realized at the very end of my clinical career that most people retire as soon as they get good at something," he told me, mentioning that his daughter likes to remind him that the university needed to hire two younger doctors to replace him because of his accumulated knowledge and diagnostic abilities.

The realization that retirement happens at the exact point when workers are at their most skilled occurred to him a year or two before he retired, when a younger colleague asked him to see a patient he couldn't quite figure out. The patient, a baby, was sitting in his mother's lap and would periodically drop and jerk his little head. "Within minutes," Jonas says, "I had an obvious diagnosis."

The baby was suffering from a seizure disorder, it turned out. Even though Jonas figured it out right away, he examined the baby awhile longer and spoke to the mother before taking his colleague aside. A light went on in the young doctor's eyes when he heard Jonas's diagnosis. By the end of the day, Jonas was the talk of the pediatric practice. As he sat down later that day to type up some notes, a second younger colleague rolled over in her computer chair and said, "Teach me, Doc! How'd you do that?"

It was something Jonas had seen before, whereas his younger colleagues simply hadn't. He realized there were probably numerous situations like this one, where older physicians were better at making diagnoses or seeing the bigger picture by sheer dint of their experience.

Though he says he wouldn't want to compete with younger doctors in trying to keep up with the latest biochemistry findings, Jonas finds it counterproductive and hurtful when hospitals and medical schools try to quietly push out older physicians. The university where he teaches now tests older doctors on cognition, regardless of whether they've given any evidence of cognitive decline, simply on the basis of their age. At least one of the doctors is fighting back by suing the university for discriminating against employees because of their age.

"When you're young," Jonas says, describing the unspoken hierarchy in clinical practice, "you tend to get brushed aside as a young whippersnapper. Whereas, once you're older, you're perceived as an old fuddy-duddy, which doesn't make any sense. It leaves only a decade or two where the respect you're given aligns with your own level of skill."

When Jonas heard about my area of research, he told me that his age beliefs have dramatically improved over time. "When I was a young pediatrician, just starting out, I thought of old people as somewhat doddering and helpless, but these stereotypes melted away as I was exposed to older mentors and peers who were thriving in their advancing age." Now that he himself is older, Jonas enjoys life with the relish of someone who finds pleasure in nearly everything he does. Yet he muses about whether there could be a role to play for retired physicians to lend a hand and share clinical experience.

Jonas works at a teaching hospital where he participates in grand rounds, which are weekly meetings where patient cases are presented to help doctors, residents, and medical students keep up-to-date on evolving areas of patient care. He shares his decades of accumulated knowledge by teaching a course on medical diagnosis to wide-eyed first-year students at a nearby medical school.

He has also thrown himself into French cooking, grows rare

orchids under hot lights, and sinks whole afternoons into his ever-expanding family genealogical tree. He is obsessed with close-up photography (he likes to focus on unexpected shapes and textures in nature). He takes long walks in the morning with a camera around his neck.

He is also a keen amateur pilot who smiles gleefully as he describes seeing and navigating the world from high up, eye level with the sun. He's been able to apply the sharp visuospatial skills he developed to diagnose and treat medical conditions to expertly scanning the skies. He told me about being in the air one recent afternoon and hearing on the radio that a search was underway for a downed small plane. He was able to find the plane, land nearby, rescue the pilot, and fly him to the wedding he'd been on his way to attend.

It helps that he has had good models of aging. His mother, for one, whom he is very close to, is ninety-seven and lives on her own in Albuquerque. As a young and then middle-aged doctor, Jonas attached himself to older mentors. From one older colleague, he learned about the vital importance of community health centers and the philosophy of community health. There was an older cardiologist who taught him a great deal about compassion and kindness toward patients. And as a pediatrician, he developed a respect for the grandparents who came in with their grandkids. "Those were the toughest visits," he tells me with a laugh. "I always felt I was under the greatest scrutiny from grandparents, grandmothers especially."

Lucky Genes, Age Beliefs, and Thriving Brains

Individuals like Jonas and Amy are living proof that our neurons and genes don't necessarily mean destiny. In fact, I found that age beliefs had a fifteen-times-greater impact than one of the best-known genes that contribute to cognition over time. Our stereo-types of aging are that powerful.[16]

Remember the *APOE* gene? The ε4 variant increases your risk for Alzheimer's; the ε2 variant, on the other hand, contributes to better cognition as we age by clearing amyloid plaques and promoting connections between the synapses of our brain.[17] In another *APOE* study, I found that people lucky enough to be born with *APOE* ε2 still benefit from assimilating positive age beliefs; they do better on cognition tests than their *APOE* ε2 counterparts with negative age beliefs. This suggests that our age beliefs can alter how our genes are programmed to influence our behavior.

The good news is that among those of us not born with the *APOE* ε2 gene (that is, about 90 percent of us), if we embrace positive age beliefs, we have the same reduced risk of developing dementia as those born with *APOE* ε2.[18] That makes sense when you remember that having positive age beliefs promotes exercise and social and intellectual engagement and diminishes stress (all of which augment brain health). In other words, age beliefs function as a sort of *cultural* ε2 variant.

For a long time, the scientific community treated the story of the aging brain as a tragedy that did not deserve investigation. It was falsely assumed that the human brain took its sweet time to develop through childhood and adolescence, then peaked sometime in early adulthood, before beginning its steady decline when its neurons stopped forming new connections. Brain science researchers have only recently begun to investigate the aging brain with the same gusto as they applied to studying the early brain.[19] They found that the neurons of older brains succeed in making new connections.

Plasticity and regeneration are qualities central to brains across the animal kingdom and across all stages of life: the brains of adult canaries are essentially "reborn" each mating season so that they can learn new mating songs[20] and stay current in the lingo of courtship and love. And you see the same kind of neural growth in older lab rats when they are given an enriched experience, such as a chance to explore interesting spaces with ramps, wheels, and toys.[21] It turns out that older brains are often regenerating.[22]

Our brains, like our other organs, must be properly cared for and nourished. In an older person with negative views of aging, who as a result doesn't exercise or stay intellectually engaged and experiences more stress, you might not see much regeneration; you might even see neuronal loss. Older adults with positive age beliefs, who might be inspired to learn to juggle, or participate in a square dancing class, or practice their high school French, might see a significant boost in neuronal growth.[23]

We are biological creatures, but we are also so much more than our biology. With the right outlook on aging, we can enhance our biological code as we age.

5

Later-Life
Mental Health Growth

I became interested in aging in a somewhat roundabout way. In high school, I spent a summer volunteering to help a psychologist who studied creativity and mental health. His office was at McLean, a Harvard-affiliated psychiatric hospital located on a beautiful tree-filled campus outside of Boston studded with converted Victorian houses. I loved the setting and was intrigued that some of my favorite poets and musicians, like Sylvia Plath, Ray Charles, and James Taylor, had all been treated there.

After college, I visited McLean's human resources office to apply for a job. I was dismayed when I heard that the only position available, since I didn't have any clinical experience, was an entry-level position that happened to be in the unit for elderly patients. I thought it would be immensely depressing. As a twenty-one-year-old, I assumed mental illness was rampant among older persons, and that these ailments couldn't be successfully treated, only managed. Those were my age beliefs back then. I pictured a sad, noisy hospital wing with helpless older patients parked in the corner or abandoned in hallways, left to their own devices. But it was my only job offer. I gave it a try.

For a year, I served meals to patients in their rooms, filled out health records, and even accompanied patients to electroconvulsive therapy (ECT) sessions. It is a procedure in which electricity is used to shock the brain, creating small seizures, which can provide relief for major depression. These treatments helped some patients who were unresponsive to other types of treatment, but I found it upsetting to watch patients convulse as electricity was administered through wires taped to their heads.

One of my favorite parts of the job, on the other hand, was writing progress notes on each of the seven patients I was assigned to monitor during my eight-hour shift. Most of my fellow mental health workers shrugged this off as busywork and often scribbled a cursory sentence or two for each of their patients: "Lisa ate most of her lunch and attended group exercise session." But I really liked the task since it gave me the opportunity to interview the patients. Whenever I spoke to one of them, I tried to learn something new about their background or how they were feeling about their family or the treatment they were undergoing. Maybe I was a little over-zealous in these progress reports, which sometimes went on for a few pages, but these notes certainly helped me better document my own learning.

During my year on the job, I learned that, contrary to my initial assumptions, mental illness is actually much less common in older than in younger adults, and that most older persons with mental illness can be successfully treated.

Almost every week, the hospital staff held team meetings to discuss each patient from a dozen different viewpoints. I watched and listened as nurses, social workers, psychiatrists, clinical psychologists, neuropsychologists, psychopharmacologists, and others squeezed inside a room in one of the hospital's elegant Victorian houses to integrate their various perspectives. For hours, they would discuss the patient's cultural background, biology, work history, and social connections to better understand what had led them to be admitted, and which approaches would help them recover.

During these meetings, I also learned that our mental health depends on the delicate interaction of many different factors. I heard, for instance, about an older Chinese woman who attributed her intense anxiety to a lack of respect from her children who refused to let her see a traditional herbalist doctor. While many other Western doctors might have disparaged the validity of traditional Chinese medicine, the medical staff at McLean did not; instead, they spent a lot of time analyzing the cultural and psychological dynamic between the patient and her children to better understand and successfully treat her condition.

Later, when I was developing my own theories and research and starting to understand how a societal factor such as age beliefs could impact and interact with biology, I would routinely think back to these enlightened meetings at McLean.

Influence of Age Beliefs on Stress

In the same way eyeglasses and telescopes transform the amount of light and detail that reaches our eyes, our age beliefs determine the kind and amount of stressors that make it through to our bodies and psyches. And these stressors, in turn, can take a toll on our mental health.

In the first study to determine how age beliefs impact our physiology, I found that positive age beliefs serve as a barrier against stress, whereas negative age beliefs amplify it.[1] I looked closely at the autonomic nervous system (ANS), which is associated with the fight-or-flight response. The ANS gives us a boost of adrenaline when we encounter a sudden threat (such as a charging bull), which immediately motivates us to combat or escape it. In the short term, this jolt of adrenaline helps us fight better or flee faster, but long-term exposure to adrenaline and stress can actually harm our health.

In this experiment, I looked at whether age stereotypes impact cardiovascular reactivity—that is, how much people experience a spike in their heart rate, blood pressure, and sweat gland activity in response to a stressor. To more closely mimic the repeated exposure to stereotypes that occur over our lifetimes, we subliminally exposed participants to either two sets of positive or negative age stereotypes and two sets of verbal and math challenges. For our verbal challenge, participants described the most stressful event of their last five years. They talked about everything from car crashes to being evicted from their apartments.

I was surprised to find that negative age stereotypes alone produced a huge and immediate dose of stress, even before the math and verbal challenges. In fact, the amount of stress was much more significant than the stress that was subsequently caused by the two sets of challenges.

The positive age stereotypes, however, had a contrasting effect. The first time we presented them, they had little impact. But the second time, they acted as a buffer: not only was there no increase in ANS stress during the second math and verbal challenges, but stress levels actually *dropped* down to where they'd been before any of the challenges. In other words, while positive age stereotypes took some time to exert their protective effect, ultimately they helped participants recover their sense of calm in stressful situations. This suggests that multiple exposures to positive age stereotypes can help older people reduce long-term stress and recover from challenging events. It also points to the interdependent nature of our age stereotypes and our physical and mental well-being.

To find out if the effect of age beliefs on stress that we found in the lab also operated over an extended period in the community, I analyzed data that had been collected over a thirty-year period by a group of National Institute on Aging researchers based in Baltimore, Maryland. Participants provided their age beliefs on their first visit to the center. Then every three years for the next three decades, when the participants returned, the researchers collected cortisol, the body's main stress hormone. As with adrenaline, limited spikes in cortisol are beneficial, whereas extensive increases in cortisol can damage the body and are associated with numerous bad outcomes.[2]

Sure enough, I found that age beliefs significantly influenced people's cortisol levels. As can be seen in Figure 3, the older participants with negative age beliefs had a 44 percent increase in cortisol during the thirty-year period, whereas those with positive age beliefs had a 10 percent *decline*.[3]

After discovering these connections between more-negative age beliefs and greater stress, I wondered whether age beliefs can also contribute to—or ward off—psychiatric conditions in later life, since stress is often a major factor in mental health issues. I undertook a study of older military veterans, whose life circumstances predictably lead to higher-than-normal rates of psychiatric conditions. After all, many have been exposed to combat, violent injury, and the deaths of fellow soldiers. In a sample of veterans from across

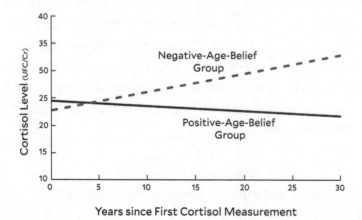

Figure 3: Stress Increases Across 30 Years for Older Participants with Negative Age Stereotypes. Participants holding these stereotypes showed an increase in the stress biomarker cortisol; whereas, those holding positive age stereotypes showed a decline in this stress biomarker.

the United States, we found that those with positive age beliefs were less likely to develop suicidal ideation, depression, and anxiety in the next four years.[4] Positive age beliefs even helped mitigate post-traumatic stress disorder among those who had seen combat.[5] In contrast, negative age beliefs made veterans less resilient in the face of adversity, as they developed higher rates of psychiatric illness.

Negative Age Beliefs as a Barrier to Mental Health

On the institutional level, negative age stereotypes held by mental health professionals can also cause harm. Providers often undertreat older patients because they think it's normal for them to have mental illness, particularly depression.[6] This makes for an especially vicious cycle: older patients with negative age beliefs have greater mental health problems, but are often unable to receive adequate treatment given the ageist nature of our health-care system, which leads to amplifying mental health problems, which, in turn, exacerbates the stereotype that these problems are inherent to aging and that older patients simply cannot be helped. Even though people over sixty-five are less likely to develop mental illness than when they were younger, one out of five experiences some sort of mental illness for which the most effective treatment is often out of reach due to ageist stereotypes that are embedded in the medical fabric.[7]

From Freud to Plotkin:
Older Patients as Particularly Treatable

The harmful stereotype that older people can't be treated for mental health problems because they are too rigid goes all the way back to Sigmund Freud, the founder of psychoanalysis, who discouraged therapists from treating older patients. He argued that among patients "near or above the age of 50 the elasticity of the mental processes on which treatment depends, is as a rule lacking—old people are no longer educable."[8] In other words, Freud held that older patients were too set in their ways to engage in the kind of self-reflection that successful therapy requires.

It is likely that some of Freud's negative age beliefs came from his ageist Austrian upbringing—particularly his own mother, Amalia. (Freud famously claimed that many of people's problems derive from their relationship with their mothers.) Freud's biographer Ernest Jones wrote about Amalia: "When she was 90, she declined the gift of a beautiful shawl, saying it would 'make her look too old.' When she was 95, six weeks before she died, her photograph appeared in the newspaper; her comment was 'A bad reproduction. It makes me look a hundred!'"[9]

Freud's beliefs about the rigidity of older patients are ironic since one of his most remarkable attributes as a thinker was that he had the honesty and courage, as he grew older, to acknowledge fundamental errors in his previous thinking.[10] In his seventies, by then world famous and nominated thirteen times for the Nobel Prize, Freud made profound revisions to his celebrated models of psychology, including about the way the unconscious drives our behaviors. But he never publicly revised his thoughts on aging.[11]

Fast-forward over a hundred years, and the situation today is not much better: Freud's ageist beliefs are still right at home in the modern American mental-health-care system. A recent survey of seven hundred psychologists and therapists found that most saw older patients as less suited to therapy due to "mental rigidity."[12] They also had low expectations for older patients' improvement (this is called "therapeutic nihilism") and many of them assumed that treatable conditions, such as lethargy and depression, were just standard features of normal aging.[13]

For a different contemporary view on mental health and aging, I spoke to Dr. Dan Plotkin, a seventy-year-old psychiatrist in Los Angeles who has found, over the course of his career, that this stereotype of older patients being rigid couldn't be more incorrect. When he started his psychoanalytic training, four decades ago, he tried to take on a seventy-three-year-old woman he called "JF," as one of the three in-depth patient case studies he was required to complete to become an analyst. He was rebuffed by his supervisors, who insisted that her age was a prohibiting factor. Analysis, they argued, only benefited younger patients who were willing to deeply search themselves and change. While reading Dan's report, "they kept angrily slashing and circling things with a red pen," Dan recalled. "They said older people can't do the deep work." Dan appealed, pointing out his supervisors' ageism, until their ruling was reversed.

This septuagenarian patient ended up benefiting in extraordinary ways from therapy. One of the issues JF was dealing with was her own aging process: she felt she had become a burden to her family because of her age. But precisely *because she was older*, Dan says, she was able to wrestle with things about herself that she hadn't been willing to tackle in earlier life. "The first time we met, she sat down and looked at me, and said, 'I want to find out what my life has been about.'"

Together, they began to unpack her life. It was such a moving first session that Dan almost started crying. Their therapy sessions continued to be productive—by the end of it, JF was the happiest she'd ever been, had made amends with her estranged daughter, and soon afterward moved to be closer to her. She was able to deal with her past and put the events of her life together in a meaningful way. At the end of her treatment, JF felt better about herself as an older person and regained her humor and creativity. "She had a renewed sense of worthiness. She was able to spend the last years of her life in the company of those she loved and who loved her."

This kind of outcome, Dan tells me, is actually quite typical of older patients' experience in therapy: they benefit just as much as younger patients. In fact, they're often more treatable than when they were younger, since they are more reflective and want to get to the bottom of things and resolve their issues. It's no surprise then that Dan prefers treating older patients.

The topic of aging comes up a lot in patient sessions, as does the idea of age beliefs. Dan says that my research helped him recognize the cultural roots of these beliefs in his patients and their impact on mental health.

A number of helpful factors make for successful therapy, he says: being motivated, being able to reflect back on one's life, and being able to form deep relationships. "Those are all characteristics we associate with normal aging!" Dan says. "In the final chapters of life, people have more maturity, a little more wisdom; they have generally found ways to be at peace with themselves. You don't have as much ego, you know your own neuroses inside out."

The science supports Dan's clinical observations that older persons may be particularly likely to benefit from therapy.[14] Studies show that in later life, we grow in emotional intelligence, spend more time on life review, have more dreams about our friends, and develop a greater respect for intuitive feelings.[15] There are exceptions, Dan says with a wry chuckle, "but when most people sit back and take a breath, in their seventh or eighth or ninth decade, looking back, most of them don't find their lives to have been a disaster."

So why do negative stereotypes about older people being inflexible and riddled with mental health problems persist?[16] In addition to long-standing biases among mental health practitioners, there are also significant structural forces that enable and reinforce medical ageism.

As with so many other problems, this one starts early, in training. Few medical schools require students to take a geriatrics course, and most geriatric courses devote at most one session to mental health. In psychiatry and psychology departments, most classes, treatments, and theories focus on childhood and young adulthood. As a result, fewer than one-third of therapists who treat older people have had any graduate training in the psychology of aging, and over two-thirds felt they needed and wanted more training in this area.[17]

Doctors are quick to give older patients medications, which require less effort and time to administer, and tend to be cheaper in the short term than combining them with psychotherapy, although many patients would prefer also including meeting with a therapist in their treatment. This approach of offering older persons a combination of medications and talk therapy has been found to be more effective than medications alone, both in terms of mental health outcomes and cost in the long term.[18] An ideal therapist, like Dan Plotkin, would identify the cause of the older patient's depressive symptoms and find a treatment that draws on and reinforces their age-specific strengths, such as increased emotional intelligence.

Misuse of medications is especially rampant in the for-profit long-term-care facilities that proliferate in many developed countries such as the United States where, in 2019, the industry was worth half a trillion dollars.[19] In these facilities, overworked staff use a number of medications to help manage symptoms of dementia, even though the Food and Drug Administration never approved many of them for this use, and they can cause fatigue, sedation, falls, and cognitive impairment.[20] In an average week, American nursing home facilities administer drugs to over 179,000 people who don't have diagnoses for which these drugs were approved.[21]

Underdiagnosing or misdiagnosing older patients' mental health problems stems in part from medical professionals' tendency to dismiss older patients' symptoms. When doctors discover suicidal ideation or depression, they're much less likely to treat it in older patients, assuming it is an inevitable feature of aging.[22] This neglect requires remedy, especially considering that older persons, particularly men, are among the most likely to die by suicide since they tend to use deadlier weapons, plan more carefully, and are less likely to be rescued in time.[23]

Bad governmental policies also explain why older people don't receive adequate mental health care. Medicare, the federal health insurance program for Americans over sixty-five, restricts older adults' already-limited access to mental health services. Medicare eligibility rules for practitioners haven't been updated since 1989, which means that even though there are approximately two hundred thousand licensed counselors and marriage and family therapists who could go a long way toward meeting some of the demand for mental health care among older patients, Medicare excludes these types of therapists from reimbursements when they treat older patients.[24] And the payments Medicare does provide other types of therapists are so low that most providers are disincentivized to treat older patients altogether. A typical psychiatrist is reimbursed for less than half of her typical fee. That's a reason why 64 percent of mental health providers do not accept older patients who rely on Medicare.[25]

But structural ageism in the field of mental health care, like our own individual negative age beliefs, is reversible. Remember my initial reluctance toward working in the geriatric unit at McLean? Dan Plotkin had a similar experience. Following medical school, he and the other medical interns were drawing straws to determine who would go first to the geriatric unit. "Nobody wanted it. This was Los Angeles; California is very youth-oriented. We were all very afraid of aging. Of course, I drew the short straw." Dan went trudging into the geriatric unit with a reluctance veering on disgust, and to everyone's surprise—most of all his own—he had a great time. He loved the staff and especially the patients. So many of them were engaged and able to talk with insight and humor about the many challenges and successes of their long lives.

The Eriksons' Vital Involvement in Mental Health

When I was in graduate school, I became friends with Erik Erikson, the émigré psychologist who coined the term "identity crisis," and his wife and frequent collaborator, Joan. Today they are best known for their theories of life span development.

I first met Erik and Joan when I volunteered as a dance teacher (I co-taught a class with a very flexible and graceful eighty-year-old ballet dancer) at the Erik and Joan Erikson Center in Cambridge, Massachusetts, a ten-minute bike ride from William James Hall, where I took my classes at Harvard. This short commute helped when I served as acting director of the center when the director went on sabbatical.

I got to know the Eriksons over meals at their nearby house, a rickety Victorian they shared with three other people of different generations. There was a young graduate student, a professional psychologist just starting her career, and a middle-aged professor of comparative religion who was always baking bread. Joan and Erik loved the lively conversations and exchanges that living communally offered them.

Erik combined Old World refinement and New World innovation. He had an elegant Continental accent and prestigious Viennese training (he studied in the same circles as Sigmund Freud and was a patient of Freud's daughter, Anna), but also an unconventional education. He trained as an artist before studying psychology and ended his formal conventional education in high school. And despite not learning English until his thirties,[26] he won the National Book Award and the Pulitzer Prize, the two most distinguished literary prizes in America.

Before Erik Erikson, theories of human development tended to focus on childhood and stop in young adulthood. Erikson, however, was interested in how social forces influence our personalities throughout the entirety of our life spans. This was partly due to his fascination with anthropology. He was a close friend of Margaret Mead, who shared his interest in how the various generations learn from one another.

In his sixties, Erik turned to Gandhi as a role model for his own development. In 1969, Erikson won the Pulitzer Prize for his psychobiography examining the later decades of Gandhi's life. Erik approached the topic not as a historian nor as a specialist on India, but as a "reviewer trained in clinical observation."[27] This allowed him to explore the historical and psychological sources of Gandhi's courage, which increased as he got older. Erikson was particularly moved by Gandhi's later-life methods of peaceful protest—such as a twenty-one-day hunger strike (his longest) undertaken when he was seventy-four to protest the British occupation.

In their eighties, Erik and Joan revised their famous psychological models of human development to include more insights into the later stages of life. This major work, based on interviews with "children of the century" (fellow octogenarians born early in the century), was called *Vital Involvement in Old Age*.[28]

"When we looked at the life cycle in our 40's, we looked to old people for wisdom," Joan said of the book. "At 80, though, we look at other 80-year-olds to see who got wise and who not. Lots of old people don't get wise, but you don't get wise unless you age."[29] Some of the participants the Eriksons interviewed for their book mentioned that humor was an important tool for coping with the unexpected. As Joan Erikson pointed out, "I can't imagine an old person who can't laugh. The world is full of ridiculous dichotomies."[30]

The Eriksons also noticed that in the eighth stage of human development, which usually starts at the age of eighty, many experience their deepest levels of intimacy. That's because, Joan said, "you have to live intimacy out over many years, with all the complications of a long-range relationship, really to understand it. Anyone can flirt around with many relationships, but commitment is crucial to intimacy. Loving better is what comes from understanding the complications of a long-term intimate bond. You learn about the value of tenderness when you grow old. You also learn in late life not to hold, to give without hanging on; to love freely, in the sense of wanting nothing in return."[31]

The psychoanalytic therapist and playwright Florida Scott-Maxwell reflects on her own experience with this stage: "Age puzzles me. I thought it was a quiet time. My 70s were interesting and fairly serene, but my 80s are passionate. I grow more intense as I age. To my own surprise I burst out with hot conviction."[32]

In the Health and Aging class I teach each year, to encourage a discussion of the inner life of older persons I show the classic Ingmar Bergman film *Wild Strawberries*. At the start of the movie, the main character, a Swedish physician named Dr. Borg, admits that he is isolated and lonely. But he is finished with self-deceptions. "At the age of seventy-six, I find I am much too old to lie to myself," he tells us. He then goes on a long car ride with his daughter-in-law, who doesn't like him, to receive an honorary degree for his fifty years of meritorious medical service. During his journey, Dr. Borg picks up travelers of different ages, representing various stages of his life. Dr. Borg also has vivid dreams throughout the course of the movie that illustrate past conflicts that he is wrestling to understand.

By the end of the film, Dr. Borg experiences a set of insights about the events and relationships in his life that allows him to connect to others in a new and fulfilling way. As a result, he wins over the admiration of his daughter-in-law as well as the young hitchhikers who serenade him outside his window before they continue their travels.

I got the idea to show the film from Erik Erikson, who told me he used to show it to the students in his life span course at Harvard, which they affectionately called From Womb to Tomb, to illustrate the stages of his theory of life span development. Both courses, his and mine, aim to show that aging can be filled with overcoming conflicts of early life, together with rewarding growth.

A Global Army of Grandmothers

Almost everything we know about improving mental health in later life is based on studies conducted in high-income countries. An important exception is a series of studies conducted by the psychiatrist Dixon Chibanda, who was born and raised and now practices in the south-central African country of Zimbabwe. He came up with an idea rooted in positive age beliefs that has transformed the mental-health-care practices of his country and improved the lives of thousands of older persons.

His idea, called the Friendship Bench, draws on the wisdom of grandmothers. The idea was born when Dr. Chibanda was working as one of only twelve psychiatrists in a country of fourteen million people. He learned that a patient of his, Erica, had taken her own life when she couldn't afford the bus fare to go to a hospital two hundred miles away. At the time, Dr. Chibanda told me, Zimbabwe was going through a period of "social, political, and economic unrest" and there was a huge gap between the availability of mental health services and the mental health needs of the population. He was determined to find a solution, but there was no funding, no space, and no available mental health workers. He couldn't even reliably enlist volunteers, since many young men and women and many older men were leaving their villages to try to find work elsewhere, often in mines.

"It suddenly dawned on me that one of the most reliable resources we have in Africa is grandmothers. Yes, grandmothers. They are in every community. And they don't leave their communities in search of greener pastures." He came up with the idea of teaching grandmothers to offer villagers talk therapy on a park bench in a safe and discreet outdoor place in the community.

At first, Dr. Chibanda "wasn't convinced it would work." He didn't know if the grandmothers would be interested or if they would have the skills needed to carry out the plan. So to find out, he recruited fourteen grandmothers without prior medical or mental health training. He taught them to administer a survey, which would establish whether the clients needed higher-level care, and to conduct the talk therapy over a series of forty-five-minute sessions.

After two months, he observed that not only were the grandmothers interested and capable of acting as lay mental health workers, but "they were actually quite talented at doing this! I realized that they were quite solid in terms of understanding what we refer to as the social determinants of mental health. They just knew what needed to be done. They actually have a lot of resources as they are, without me even coming on board. My work was really to empower them with a structured way of using the tools and knowledge that they already have." Among these tools, they "are good listeners, have an ability to convey empathy and reflect, and they draw on local wisdom and culture."

One reason the Friendship Bench has caught on in Zimbabwe is the culture's strong positive age beliefs. Dr. Chibanda pointed out that "the first thing that sticks out in my culture when it comes to aging is respect. There is a lot of respect for older members. I think that's probably why the Friendship Bench is successful. The clients acknowledge that the grandmothers have one or two things to teach you."

Today there are eight hundred grandmothers, with an average

age of sixty-seven, who offer talk therapy to their fellow villagers. The Friendship Bench model, which has been extended to Malawi, Botswana, and Zanzibar, has led to the treatment of over seventy thousand clients of all ages. When the patients are younger, the grandmothers call them "grandchildren." When the clients are closer to the grandmothers' ages, they call them "brother" or "sister."

The success of the program has been documented in several clinical trials. In one, published in the prestigious medical journal *JAMA*, Dr. Chibanda's team found that the "grandmothers were more effective than doctors in reducing depression."[33] In another trial, they found that the grandmothers themselves also benefited from providing therapy on the Friendship Bench. On the one hand, Dr. Chibanda finds this surprising because the grandmothers are "spending so much time talking to people who are traumatized." On the other hand, he explained it makes sense to him: "This work gives them a sense of belonging and purpose. They do much better than other elders who are not doing this work because they are giving back to the community—the community that has taken care of them all these years. And now, in the twilight years, they are giving something back to the community and feeling this immense reward."

Grandmother Kusi, an eighty-year-old woman living in Mbare, is one of the grandmothers that Dr. Chibanda particularly admires. She was one of the original fourteen grandmothers first recruited fifteen years ago and has by now successfully treated hundreds of clients on the Friendship Bench. He explains, "What makes her one of the most effective grandmothers is this amazing ability to give people space to share their stories. She's also an unbelievable storyteller herself. She knows exactly how to use her body language— the way she communicates with her hands, with her eyes. She is great at listening and knows when to give someone a hand if they are crying; all those little things that you're not taught at medical school or in psychiatry. She's just brilliant."

Dr. Chibanda has a dream: "There are 1.5 billion people aged sixty-five and above. Imagine if we could create a global network of grandmothers in every city!" This army of grandmothers (it could also include older women without children and older men with and without children, he believes) would be able to provide mental health services to the millions of people in need who do not currently receive any treatment.

Even though the Friendship Bench model benefits from a culture that already has positive age beliefs, it has also succeeded in pilot projects in other countries that have more-negative age beliefs. It seems the example of the grandmothers providing effective talk therapy could help upend these negative beliefs. When I heard about Dr. Chibanda's mental health vision, I asked him whether he thought the Friendship Bench could also contribute to a reduction of ageism. He agreed this could also be part of the dream.

6

Longevity Advantage
of 7.5 Years

A few decades ago, a team of researchers descended on the small, quiet town of Oxford, Ohio, and invited every resident over the age of fifty to join a project called the Ohio Longitudinal Study on Aging and Retirement. They asked these Ohioans a slew of questions about their health, work life, family—and their thoughts on aging. The latter included questions such as "Do you agree or disagree that as you get older you are less useful?"

Suzanne Kunkel, who today leads the Scripps Gerontology Center at Miami University, first moved to Ohio right out of college to join the research team. She was a new graduate student in sociology with an interest in human development. Robert Atchley, the study's director, wanted to recruit as many residents as possible, so Suzanne spent her first few weeks in Oxford taping fliers on the doors of restaurants and coffee shops, scouring voter rolls, sending postcards to everybody in town asking community members to search their Rolodexes, and encouraging everyone to spread the word. The goal was to include all residents of the right age, whether they were retired surgeons or car mechanics, whether they lived in one of the stately brick homes on Church Street or in a trailer, whether they were Scottish or Laotian, or conservative or liberal. Atchley felt these differences would help his analysis and allow him to isolate the role of sociological factors on aging.[1]

Over the next few decades, Suzanne and her fellow investigators returned five more times to ask follow-up questions. The resulting study would provide one of the richest and most detailed perspectives on aging in late-twentieth-century America, yet some of its profoundest implications sat uninvestigated and ignored for a quarter century. I stumbled upon it in graduate school soon after I returned from Japan. Having just spent months in a place where centenarians were common and old age wasn't shunned but celebrated, longevity was on my mind. At that point, I suspected culture played a major role in shaping people's age beliefs and I wanted to know if age beliefs, in turn, might have a demonstrable impact on longevity. I heard that the Ohio Longitudinal Study had an age-belief measure collected at baseline.

When I contacted Suzanne Kunkel and pitched her my idea, she told me that although a number of participants had died over the years, their longevity had never been recorded. Accordingly, there was no way to know which members of the study were still living or dead.

By a stroke of luck, soon afterward, I attended a conference on aging where I would discover a way to fill this gap. As I strolled through the exhibit hall with a tote bag bulging with longevity-related swag (a beach towel, Frisbee, and a visor—that became some of my unusual beach gear), a friendly man in a polka-dot bow tie handed me a ruler with "NDI" imprinted in big, bold letters. Puzzled, I asked him what NDI meant and he explained that it stood for the National Death Index, a government effort to keep track of the longevity of all Americans. Like the birth registry, he said, but at the other end of life. "Perfect!" I shouted, startling more than a few people in that quiet exhibition hall.

The conference was teeming with famous longevity experts. Everyone was studying the topic from a different perspective: one person was using fruit flies as his approach; someone else was studying the blood pressure of centenarians; a third was looking at demographic trends in Sweden. No one, however, seemed interested in psychological determinants, such as age beliefs.

Now I had a way to examine whether this link exists by overlaying age beliefs onto this new mortality data. I looked at the Ohio Study participants' age beliefs starting in middle age and tracked them over time. What I found was startling. Participants with the most-positive views of aging were living, on average, seven and a half years longer than those with the most-negative views.[2]

Since there was so much information collected about these Ohioans, I was able to determine that age beliefs were determining their life spans above and beyond the influence of gender, race, socioeconomic status, age, loneliness, and health. Age beliefs stole or added almost eight years to their lives, conferring an even better survival advantage than low cholesterol or low blood pressure (both of which added an extra four years of life) or low body mass index (one extra year) or avoiding smoking (three extra years). (See Figure 4.)

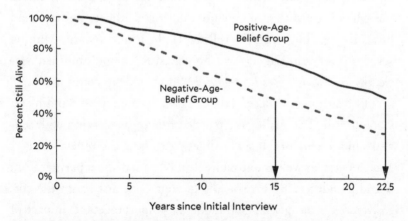

Figure 4: Survival Advantage from Positive Age Beliefs. Participants with positive age beliefs lived an average of 7.5 years longer than those with negative age beliefs. This was calculated by examining the group difference in the time it took for half the people to be still living, as indicated by arrows.

In my article about these findings, I concluded, "If a previously unidentified virus was found to diminish life by over 7 years, considerable effort would probably be devoted to identifying the cause and implementing a remedy. In the present case, one of the likely causes is known: societally sanctioned denigration of the aged. A comprehensive remedy requires that the denigrating views and actions against elderly targets undergo delegitimization by the same society that has generated them."[3]

The study received so much press that for a few days, my life gained a surreal quality. I went from spending my days alone in quiet concentration, reading and writing in a cubicle tucked underground in a Gothic library at Yale, to being chased down the street by local, national, and foreign radio, print, and TV journalists. It was shocking to suddenly become the center of attention, but I was glad that ageism and age beliefs were getting recognition.

A few weeks after the study was published, I received a phone call from Washington, D.C. It was a staffer for US senator John Breaux, who wanted me to share my new findings at a hearing on ageism. I hesitated, already reeling from all the sudden attention, but when I heard that my mentor and friend Robert Butler would be testifying, along with Doris Roberts, the Emmy Award–winning seventy-seven-year-old actress who played Raymond's mother, Marie, on *Everybody Loves Raymond*, I told the aide to count me in.

The hearing took place in the Dirksen Senate Office Building on Capitol Hill. The mahogany-paneled room was teeming with senators and journalists. It was validating to hear the others—Butler and Roberts, as well as journalist Paul Kleyman and a partner from a leading advertising firm—talk at length not just about the corrosive effects of ageism, but about the importance and impact of age-related imagery. Butler showed two images. One was an illustrated magazine cover showing a horde of stooped-over, frowning older men and women (titled "Greedy Geezers"); the other was an elegant, glamorous, even sexy portrait of ninety-two-year old Kitty Carlisle Hart, an actress and opera singer (from a collection of photographs called *Wise Women*). Butler's point in showing these two contrasting images was that media and marketing companies didn't need to denigrate older people; there were many other ways to present them.[4]

Then, Doris Roberts spoke about the personal impact of these portrayals. "I am in my seventies, at the peak of my career, at the height of my earned income, and my tax contributions, I might add. When my grandchildren say that I rock, they are not talking about a chair. Yet society considers me discardable. My peers and I are portrayed as dependent, helpless, unproductive, and demanding rather than deserving." Notice the word she used: "portrayed." Doris Roberts, too, focused on the issue of representation and imagery.

"In reality," she went on, "the majority of seniors are self-sufficient middle-class consumers with more assets than most young people and the time and talent to offer society. This is not just a sad situation, Mr. Chairman. This is a crime. . . . The later years can be some of life's most productive and creative. For the last 100 years, the average age of the Nobel Prize winner is 65. Frank Gehry designed Seattle's hip new rock museum at the age of 70. Georgia O'Keeffe was productive way into her eighties. Add to the list Hitchcock, Dickens, Bernstein, Fosse, Wright, Matisse, Picasso, and Einstein, just to mention a few people who produced some of their best work when they would be considered over the hill by current standards."

As an actress, Roberts reserved the brunt of her ire for the image makers—the entertainment industry she belonged to and worked in. Even though she noted that actresses get "better and better in their craft as they get older," she added that many of her friends, talented actresses in the forty-to-sixty-year-old range, were "forced to live on unemployment or welfare because of the scarcity of roles for women in that age bracket."[5]

When it was my turn to address the senators, I explained how I discovered that negative age beliefs affect not only health outcomes, such as memory performance and cardiovascular response to stress, but also the very length of our lives.

People I meet still bring up this longevity finding: "Oh, you're the person who discovered that finding about the seven and a half extra years of life!" Since I published the study, its findings have been replicated in ten countries, in places as different from one another as Australia, China, and Germany, and have become the cornerstone of a recent World Health Organization campaign to combat ageism.[6]

A group of older activists in Wisconsin sent me some of the buttons they made, based on our longevity finding, as a way to start conversations with people about the power of age beliefs and the need to fight ageism. The buttons say, "ASK ME ABOUT 7.5." The reason this finding resonates with so many is that it refutes the widely held assumption that genes alone determine how long you live. Knowing that age beliefs play such an important role in our longevity means we can increase control over our life expectancy. So it's easy to understand why people would wear an "ASK ME ABOUT 7.5" pin, rather than an "ASK ME ABOUT GENES" pin.

In fact, research shows that nonbiological factors, including age beliefs, determine as much as 75 percent of our longevity.[7] And the 25 percent of our longevity thought to be determined strictly by our genes may be even lower still, considering that we found age beliefs impact whether, and if so how, these genes are expressed.[8] Yet most of the recent and ongoing research into the determinants of longevity has focused on genes.[9] What's more, most of these investigations of nongenetic factors focus on negative ones, such as disease, injury, and cognitive decline, rather than positive factors, such as protective age beliefs.

Genes are powerful, for sure, but so is the environment. Some centenarians are born with lucky genes (like *APOE* ε2) that get passed down from generation to generation. But many centenarians have no such genes.[10] Some even have the risky *APOE* ε4 gene but are able to overcome it with their environment, which includes how they are treated by others and the beliefs they assimilate from their environments.

Take the queen honeybee's life span as an analogy of how the environment can overcome genes. Queen bees and worker bees share identical genes, yet the queen lives five times longer than the workers. Although they share a beehive, they essentially live in two different environments. Instead of the usual pollen eaten by the workers, the queen is constantly groomed and fed a special royal jelly by a whole "court" of attendants who predigest her food.[11] In other words, when it comes to factors that determine longevity, the social environment can trump genes. One such social environment for humans, a culture rich in positive age beliefs, not only unites the generations, it also can extend survival regardless of one's genetic script.

Pathways to Survival

So how is it that the age beliefs we take in from our social environment have anything to do with survival? We know from my stereotype embodiment theory, which explains how these beliefs impact our health in later life, that there could be psychological, biological, and behavioral pathways.[12]

The psychological mechanism includes will to live, which is the feeling that the perceived benefits of life outweigh its perceived hardships. If that's a little abstract, think of it as something to look forward to. Costa Ricans talk of "plan de vida." Americans say "why I get up in the morning." The French have "raison d'être." And the

Japanese say "*ikigai*." These can all be translated as "will to live" or "reasons for being."

Will to live isn't a lofty philosophical conviction, but simply the sense that life is worth living. We manifest it when we care for a loved one, look after a pet, tend a garden, or engage in work that contributes to society. Things that give us purpose, the feeling that we are useful, generate will to live.

My late colleague and epidemiologist Stan Kasl demonstrated that will to live, or in this case even just having an event to look forward to, can extend life. He and sociologist Ellen Idler found that observant Christians frequently delay their deaths until after Christmas and Easter, whereas observant Jews often delay their deaths until after Yom Kippur, Passover, and the Jewish New Year.[13] When I talk about this phenomenon in my Health and Aging class, there are always a few students who raise their hand to share stories of relatives who delayed their death until after a much-anticipated wedding or birth.

I was able to demonstrate the dramatic effect of age beliefs on older persons' will to live in an experiment where half the participants were primed with positive age stereotypes and the other half with negative ones. Then, all were presented with the following scenario: What if you developed a fast-acting illness that was sure to kill you within a month unless you opted for an aggressive and expensive treatment? Let's say this treatment guaranteed a 75 percent chance of survival, but it would cost almost all your savings, and on top of that, your family would need to spend many hours caring for you. We found that younger participants, regardless of whether they'd been implicitly exposed to positive or negative age stereotypes, tended to accept the life-prolonging treatment. In contrast, older participants, for whom the age stereotypes were self-relevant, tended to accept or reject this life-prolonging treatment based on whether they'd been exposed to positive or negative age stereotypes, respectively.[14]

As examples, Ernie, a sixty-four-year-old Bostonian who owned a concession stand at Fenway Park, decided he would rather die than accept the life-extending treatment. He'd been exposed to negative age beliefs. On the other hand, Bette, a sixty-five-year-old who ran a Jamaica Plains hair salon and had been exposed to positive age beliefs, said she would definitely opt for the treatment.

In our Ohio study, we were able to show that the will to live is one of the ways age beliefs impact survival. All our participants filled out a will-to-live measure. Those with negative age beliefs were more likely to describe their lives as "worthless" or "empty," whereas those with positive age beliefs were more likely to describe their lives as "worthy" or "full." Among our Ohio participants, those who managed to ward off negative age stereotypes expressed a stronger will to live and this, in turn, predicted longer survival.[15]

Although it hadn't been previously shown, I had a hunch that the biological pathway by which age beliefs impact longevity involves the way we experience stress. I took a closer look at a stress biomarker called C-reactive protein (CRP), which is a ring-shaped protein found in blood plasma that rises in reaction to cumulative stress.[16] People who die earlier generally have higher CRP levels.[17] We followed more than four thousand Americans aged fifty and older for six years, tracking their age beliefs and CRP levels. It turns out that positive age beliefs predicted lower CRP, which led to longer-term survival. That is, positive age beliefs increased the ability to resist and cope with stress at a biological level, which impacted longevity.

Finally, there is a behavioral dimension that links age beliefs and survival. It's how people approach their health care. A common theme of negative age beliefs is that debilitation in later life is inevitable. As a result, we found that people with negative age beliefs, compared to those with positive age beliefs, are less likely to engage in healthy behavior, since they regard it as futile.[18] I explored the reach of this finding during the early stage of the COVID-19 pandemic, when the US was in lockdown. My team measured the age beliefs of 1,590 older and younger participants and asked them whether they thought older people who were very sick with COVID-19 should go to the hospital for treatment or forgo treatment by staying home.[19] Younger participants' age beliefs didn't affect their answers because these beliefs weren't self-relevant. As for our older participants: the more negative their age beliefs, the greater their resistance to hospitalization. This was likely because of the futility they felt about treatment. In contrast, those with more-positive age beliefs tended to favor older persons getting needed treatment in hospitals.

Living the Dream:
Meeting Our Great-Great-Grandchildren

Longevity is not a new dream of humankind. As the historian Thomas Cole points out, "People of all times and places have dreamed of longer life, if not immortality."[20] We can find examples throughout history. Xuanyuan Huangdi, the emperor who founded Chinese culture five thousand years ago, long sought immortality. The ancient Greeks believed in gods who ate ambrosia to fend off death. In Goethe's classic German tale *Faust*, the protagonist bargains with the devil for immortality. Peter Pan never grows old. More recently, there is the immortal teenage-heartthrob/vampire Edward Cullen, one of the main characters in *Twilight*, the wildly popular series of romance fantasy novels and movies.

Given our enduring fascination with living as long as possible, one might think that the widespread increase in longevity that our species has achieved would be cause for a correspondingly widespread celebration. This is far from the case, as we will later show. Our life spans have effectively tripled from what was the normal life span throughout most of human history.[21] In the last 120 years, we have added thirty years to life expectancy. As Robert Butler points out, "In fewer than 100 years, human beings made greater gains in life expectancy than the preceding 50 centuries."[22]

And there is no sign that this trend is leveling. Our steady increase in life span is in fact one of the most linear and consistent trends ever observed in nature. As demographers James Oeppen and James Vaupel point out, "In the last 160 years, there has been a steady rise of 3 months per year of life expectancy."[23]

Of course, there are variations in life expectancy by region, gender, and ethnic groups. People in high-income countries and with greater resources generally live longer than those in low-income countries, and women everywhere tend to outlive men. But these trends are not always in ways you might expect. In the US, for instance, although a number of factors, including structural racism, have led to African Americans having shorter average life expectancies than whites for younger age groups, this trend reverses for those who make it past eighty, with Black elders living on average longer than white elders.[24] One reason that Black elders may have a later-life survival advantage is that several studies have found that compared to the white culture, the Black culture has more-positive age beliefs. This may be rooted in the culture having more intergenerational households with grandparents often involved in childcare; it is known that intergenerational contacts foster more-positive age beliefs for both generations.[25]

Instead of viewing the global increase in longevity as the victory that humanity has dreamt of for thousands of years, it is largely portrayed as a natural disaster that will burden world populations.[26] Since the 1980s, policymakers, journalists, and commentators have blamed countless economic problems on the expanding older population and raised alarms about impending national bankruptcies, when the real culprit was often rapidly growing economic disparities, with wealth concentrating generation by generation in fewer and fewer hands.[27] (A term was recently coined—"centibillionaires"—to describe the new financial category, which includes Jeff Bezos and Elon Musk, whose net worth each topped one hundred billion.[28])

Health and Wealth with Longevity

Even though there is a perception commonly presented in the media that increasing longevity will sap the public coffers and overfill our hospitals, growing evidence shows that increasing longevity is actually a harbinger of health and wealth. As Linda Fried, dean of Columbia's Mailman School of Public Health, aptly put it, "The only natural resource that's actually increasing is the social capital of millions of more healthy and well-educated adults."[29] "Social capital" is a broad term that refers to societal contributions generally, but the resources brought by longevity also include capital in the traditional financial sense. A study of thirty-three wealthy countries found that population aging correlated *negatively* with health expenditures.[30] In other words, the older the population gets, the less its country will need to spend on health care.

Additionally, growing life spans lead to what Joseph Coughlin, of the MIT AgeLab, calls the "longevity economy."[31] Those over the age of fifty control 77 percent of the total net worth of US households and spend more on travel, recreation, and personal-care products than any other age group, even though they only make up 32 percent of the population.[32]

Far from being a drain on the economy, as negative age stereotypes misconstrue, older people are helping to drive it: the collective flow of private funds within families is much higher from older to younger relatives than the other way around.[33] When you look at entrepreneurs, who are celebrated in the US as modern heroes for creating new businesses and jobs, the successful ones are twice as likely to be over fifty than in their early twenties.[34] And economists have found that for many countries increases in longevity lead to increases in gross domestic product.[35] In Singapore, older parents often live with their poorest adult child so that they can best distribute their support. Researchers of this phenomenon found the older parents "cited the desire to experience the psychological gratification and the appreciation they derived from providing material support along with love and companionship."[36]

Contrary to the ageist myth that growing life spans are a catastrophe for our health systems, longevity actually provides significant health dividends. The myth is based on the pernicious stereotype that aging brings with it a range of inevitable physical and mental illnesses, accompanied by ballooning medical costs. But accumulating evidence suggests that as humans live longer, we are experiencing what James Fries, who teaches medicine at Stanford, calls a "compression of morbidity," or a growing number of years free of illness.[37] Common illnesses like heart disease and arthritis occur increasingly later in life. Today, people are two and a half times likelier than a century ago to enter their sixties without any chronic illnesses.[38] Older people are healthier and more vigorous than ever before, and disability and disease rates are going down.[39]

Thomas Perls, who started and now oversees the world's largest study of centenarians and their families, the New England Centenarian Study, is a friend from my postgraduate days at Harvard, when we shared an office suite in an old brick building above one of my favorite diners. We bonded quickly when I learned that he decided to become a geriatrician after reading Robert Butler's seminal book on aging and ageism, *Why Survive? Being Old in America*.

In his groundbreaking work, Tom discovered a pattern that he hopes will refute the ageist idea that the older you get, the sicker you get. Instead, what he has found is that "the older you get, the healthier you have been." He explains, "That's what we're seeing with the centenarians. To live to older ages, you can't have been sick for a period of time. You have to age slowly or escape the age-related diseases."[40]

In one study, Tom found that 90 percent of centenarians were functionally independent in their nineties—meaning they lived their days without any help whatsoever.[41] "The centenarians I have met with have with few exceptions reported that their 90s were essentially problem-free. As nonagenarians, many were employed, sexually active and enjoyed the outdoors and the arts."[42] Most supercentenarians (those over the age of 110) were living on their own at 100, and few had diabetes or vascular-related diseases, including hypertension.[43] Similarly, in a recent study 330 Dutch centenarians demonstrated preserved abilities in a range of cognitive tasks including listing animals beginning with a certain letter and not getting distracted when working toward a goal.[44]

Tom thinks that studying centenarians "will yield clues not so much about how to get people to extreme ages, but how to help them avoid or delay diseases like Alzheimer's, strokes, heart diseases, and cancer."[45] People who live *long* lives, in other words, can teach us how to live *healthy* lives.

The Secrets of Extreme Longevity

Japan, as we've seen, has discovered the secret to longevity. Both Japanese men and women enjoy the longest life expectancies in the world, and more centenarians and supercentenarians live in Japan than anywhere else.[46]

The world's oldest living person is a Japanese woman named Kane Tanaka, born the same year the first airplane took flight. Today she is 118 and lives in Fukuoka, on the northern shore of Kyushu, an island in the Okinawa region.

Perhaps it's no surprise that the name *Okinawa* includes the kanji, or Japanese symbol, for *okina*, which means "old man," or "venerable," since old age is something Japanese culture venerates. When people turn 61, 77, 88, 90, 99, 100, and the aspirational 120, they receive special gifts.[47] (Contrast that with the gallows humor or bleak birthday cards that often accompany later-life birthdays in the West.) On *Keiro No Hi*, Japan's annual celebration of the aged, the government sends all centenarians and supercentenarians checks, and each prefecture holds a party for its oldest citizen.

During one recent such party, the mayor of Kane's town brought her a huge cake in the shape of her favorite board game, Othello. Even though he knew how competitive she was, he challenged her to a game. They used the cake as their board as a stunt for the camera crews, which the mayor knew would save him from a rematch; this was clever, since Kane hates losing so much that she often challenges her opponents to rematch upon rematch until she wins. Later, when Kane became the oldest living person, the same mayor attended the ceremony and watched her bow low to the ground and tell the crowd that she was now the happiest she's ever been.[48]

That's the life of a supercentenarian in Japan: you're treated like a pop star. Kane often goes on Japanese TV and recently appeared in a period drama, as well as a reality TV show that features a trio of celebrity guests each week (Kane shared screen time with a famous comic and a popular model). She also appeared with her great-granddaughter on a show that highlights inspiring, real-life Japanese stories.

These days, Kane fills her time with calligraphy, journal writing, origami, and board games. She also belongs to a math club that meets daily to work on math challenges and stays physically active.[49]

What is it about Japan that promotes such long lives? When I spoke to Yumi Yamamoto, who works for an organization that verifies the life spans of Japan's oldest living people, she told me about her great-grandmother and role model, Shigeyo Nakachi, who at the time was the fifth-oldest person in the world, at 115. From interviewing Japanese supercentenarians Yumi has noticed that, like her great-grandmother, they all share positive attitudes about aging and a deep gratitude for their families' appreciation and respect.

Yumi's boss, an American named Robert Young, has been verifying the ages of the world's oldest living people for the *Guinness Book of World Records* for the last sixteen years. It's a busy job that requires a lot of careful detective work throughout the world. Because humans aren't like trees, with rings to mark our years, he spends his days verifying life spans by tracking down ancient photo IDs and birth records and musty marriage licenses. When I asked Robert what it was about Japan that promotes longevity, he smiled as if he had been expecting the question. "Culture. It's culture." He went on to tell me about the country's Confucian roots, which have for centuries promoted deep respect toward the nation's oldest members and a widespread appreciation for their good advice and hard-earned perspective.

We have found in cultures with positive age beliefs that they usually extend from top to bottom. In Japan, it's not only the very old who feel good about aging. Japanese children, too, are taught to enjoy and look forward to spending time with their elders. Grandchildren often live with or near their grandparents, with whom they often share a special bond, and many characters in folktales for children are older people who give off a sense of infectious happiness and contentment. The *jiisan* and *baasan*, grandpas and grandmas, in these stories are portrayed as kind, healthy people, and these tales usually have a happy ending.[50] (Contrast this with some of the popular children's tales that feature older characters in the US and Europe, such as "Hansel and Gretel," whose antagonist is an older witch who attempts to roast and eat the children.)

Even though Japan has evolved and modernized with the rest of the world, because it had been a closed society it maintains many more elements of its traditional culture than more heterogeneous countries like the US or Canada. This traditional culture, in turn, wields a profound influence on the ways Japanese people think and live.

Japanese culture is "collectivistic," which means that Japanese individuals are seen as interdependent and embedded within a larger society. On the other hand, "individualistic" cultures, such as the US, valorize the autonomy and independence of the societal members.[51]

Cultural psychologists Hazel Markus and Shinobu Kitayama dramatize this cultural chasm with a parenting example: "American parents who are trying to induce their children to eat their suppers are fond of saying 'Think of the starving kids in Ethiopia and appreciate how lucky you are to be different from them.' Japanese parents are likely to say 'Think about the farmer who worked so hard to produce this rice for you; if you don't eat it, he will feel bad, for his efforts will have been in vain.'"[52] Or consider the way Japanese and American corporations seek to motivate their employees. A Texas corporation seeking to elevate productivity tells its employees to look in the mirror and say "I am beautiful" a hundred times before coming to work each day. In contrast, the employees of a Japanese-owned supermarket in New Jersey are instructed to begin the day by holding hands and telling each other "You are beautiful."[53] One culture views the self as primarily an individual, whereas the other sees the self as part of a larger network.

This interdependence, in turn, promotes and supports a culture of positive age beliefs. In their study that looked at a million people in sixty-eight different countries, William Chopik and Lindsay Ackerman found that members of collectivist cultures express less explicit and implicit ageism, and a greater respect for elders.[54] We know from our research that these positive beliefs, in turn, predict a longer life span.[55]

Longevity Equation

So what do these observations about culture tell us about the age code as it applies to longevity? To think about how the different components intersect, I want to propose a longevity equation:

$$L = f(P, E)$$

In this equation, longevity (L) is a function (f) of both the person (P), which includes personality and genes, and the environment (E), which includes the physical and social surroundings. Age beliefs start in the environment, but then become absorbed by the person. In other words, they're a joint effort between individuals and the culture they live in.[56] This environment can convey an appreciation for older persons, or in the case of structural ageism, it can convey a stigmatization of older persons.

Robert Young, the *Guinness Book of World Records* chief longevity expert, was born in Florida, but as a child he shared Japan's profound affection for old people. When he was three years old, his favorite great-uncle died, "because he was old," his mother told him. "At that point, I decided I would become friends with old people first because they would die first." A year later, at four years old, Robert remembers having what he calls a "wow moment" when he saw a 108-year-old woman on the local news. He wondered how she'd been able to live so much longer than his uncle, and he later became fascinated with life span patterns throughout the world. As a teen, he clipped and collected articles about super-agers and began to write to the *Guinness Book of World Records* with suggestions about who he thought might be the world's longest-living person.

But it was more than just a passion for documenting the lives of supercentenarians. These people, Robert soon realized, were living links to the historical past, with stories and humor and ways of seeing the world that were increasingly rare—and valuable. Early on in his career, for instance, he met 115-year-old Betty Wilson in

Mississippi and saw history come alive. Painfully, poignantly, she told Robert about growing up in the Reconstruction South, the daughter of slaves. She told him about teaching herself to read and write, about the atrocities of the Jim Crow era, about the resilience and hopefulness that carried her through her life. She handed him her cane, its handle smoothed to a high gloss by a century of warm hands, and explained it had been carved by her enslaved relatives. Every day, her ancestors helped her walk through the world.

Positive age beliefs have a double benefit for longevity. In addition to the likelihood of a longer life, the various rewards these beliefs provide make it more likely the longer life will be a fulfilling and creative one.

7

Stars Invisible by Day: Creativity and the Senses

One Size Does Not Fit All

Not long ago, my younger daughter returned home from college for a long weekend, excited to talk to us about her new chosen fields of study. She had decided to major in philosophy and cognitive science and was glowing with the excited fervor of a new convert. To explain how these fields inform our view of the world, at dinner, she grabbed a marker and a napkin and drew two different daisy shapes, one with six large petals and the other with an equal number of smaller petals. "Which of the two central circles is bigger?" she asked.

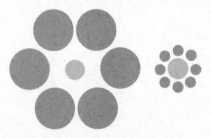

Without hesitating, I pointed to the drawing on the right, since its central circle looked twice as big as the other one. My daughter smiled as she took a napkin, drew dashes to measure the diameter of the central circles, and then folded the napkin in half so the two sets of dashes were next to each other. The dashes were the same length. In other words, the circles were the exact same size.

Maybe you are familiar with this optical illusion. I wasn't, but I can now be counted in the long line of people who have been tricked by it for over two hundred years, since it was first demonstrated by the German psychologist Hermann Ebbinghaus. It's still used today as a way to illustrate the pitfalls that lie in wait for our brains as we process information about our surroundings.

What I like about this particular illusion is that it demonstrates how our perception is impacted by its context. In this case, the size of the petals influences how we perceive the circle in their center. Another interesting aspect of this illusion is that children tend to be immune to it. In other words, the illusion is a perceptual defect that we pick up as we navigate the world through time, and it's especially strong in those adults who are most sensitive to their context or environment.[1]

There is a small rebellious subfield of social psychology called the New Look, which studies how our perceptions of objects and events are often influenced by unseen social and cultural forces. It was developed by the psychologist Jerome Bruner, who was born blind, with cataracts in both eyes. His vision wasn't restored until he was two.[2] He spent the rest of his lifetime trying to understand how we perceive the world. In one of his famous studies, he showed that children from poor backgrounds perceived the size of coins as much larger than did children from wealthier backgrounds.[3] This turned conventional sensory and perception research on its head, since it was long assumed, and still often is today, that we process the world in a more or less objective fashion. But it turns out that our beliefs and experiences come to bear not just on who we are, but literally on how we see the world.

Solomon Asch, in a now-famous set of experiments, found that our perception of the size and length of objects can be influenced by peer pressure and our desire to conform. In a follow-up study using brain scans, it was discovered that when people make judgments under social pressure, it's not just that they pretend to see things differently to "fit in," but rather that social pressure transforms the part of the brain that perceives the size and length of objects.[4]

I found that age beliefs can also impact perceptions. After participants were implicitly primed with age stereotypes, they read a short description of a fictional seventy-three-year-old woman who imagines that a person and an object (a crumpled napkin) look like animals. Those exposed to negative age stereotypes tended to see her daydream as an indication of dementia; those exposed to positive age stereotypes tended to see it as an indication of creativity.[5]

Is it possible that age beliefs influence not only our perceptions but also our ability to use our sensory systems, such as our hearing, and our creative processes? The rest of this chapter will explore these connections.

Hearing Culture

In the 1980s, a Chilean ear surgeon named Dr. Marcos Goycoolea traveled to Easter Island, famous for its mysterious *Moai* stone sculptures of gigantic heads, to investigate another mystery: the hearing of the island's oldest residents. He found that those who had spent their lives on the island were able to hear much better than those who had lived for a time on the South American mainland (Chile annexed Easter Island in 1888, and there is a fair amount of emigration from the island to the mainland). Goycoolea thought the discrepancy might be due to the contrasting levels of everyday noise: Easter Island was remote and quiet, smack-dab in the middle of the Pacific, while Chile was urbanizing rapidly, full of the noises of machinery, honking cars, and the other din of city life.[6]

When I read his findings, though, it occurred to me that there might be a different explanation for this hearing advantage. Was it possible that age beliefs played a role? As I combed through anthropological texts, I discovered that Pacific Islanders traditionally held positive age beliefs, while South American age beliefs were becoming increasingly negative.[7]

To test my theory, we interviewed more than five hundred older people from the New Haven area.[8] Our nurses fanned out to visit participants in their homes, armed with handheld audioscopes that emitted a range of beeps (representing the range of tones used in everyday speech) inside the ear canal. Participants were asked to raise a hand when they heard a beep. We found that older people who held positive age beliefs at the start of the study heard more of the beeps over the next three years than those who held negative age beliefs at baseline. In fact, those with the most-negative age beliefs had a 12 percent greater hearing decline over the next three years than those with the most positive age beliefs. They were a greater predictor of hearing than other known factors, such as smoking. Later-life changes in sensory perception are often assumed to be determined solely by our biology, but here was evidence that these changes are also influenced by culture.

Other researchers have since come to the same conclusion.[9] In one study, psychologist Sarah Barber randomly assigned older participants to read either of two stories.[10] In one, they read that younger people have been losing their hearing by listening to loud music on their headphones. In a second (false) story, they read that all older people lose their hearing. Many more of the people who read that second story subsequently reported hearing problems, compared to those who read the first story.

Listening to the Old-Time Greats

It's not just the Polynesian culture of Easter Island that highlights the association of age stereotypes with sound. Popular music is replete with fear of aging (cue the Rolling Stones' "What a drag it is getting old," Miranda Lambert's "Gravity Is a Bitch," and the Who's "I hope I die before I get old"). But there are numerous musical subcultures that espouse more-positive age beliefs.[11] That's why it's common to see older musicians, such as jazz greats Sonny Rollins and Allen Toussaint, take the stage well into their seventies and eighties. In fact, the field of music is filled with performers who seem to get better as they grow older. Take 82-year-old singer-songwriter Mavis Staples, who released a record number of popular albums in later life, or composer Elliott Carter, who experienced a renewed burst of creativity at age 90 (when he composed his first opera), which lasted until his death at 103. Or one of my personal favorites, Leonard Cohen, whose chillingly beautiful final album, *You Want It Darker*, was released when he was 82 and spent weeks atop the charts.

With so many examples of later-life success, many musicians don't adhere to the narrative of aging as a trajectory of sensory and cognitive decline. This may be why they can actually hear better than nonmusicians later in life. Older musicians are 40 percent better than nonmusicians at hearing speech in noisy environments (say, a really loud restaurant), and the average seventy-year-old musician hears about as well as the average fifty-year-old nonmusician.[12]

This old-age hearing advantage likely has a link to age beliefs. Knowing many older role models in their field contributes to musicians often continuing to devote themselves to performing as they grow older. Musical experience, or any serious pursuit spent interacting with sound in an active manner, such as birding, engages those parts of the brain devoted to interpreting and making meaning out of sound, which improves hearing,[13] which in turn reinforces positive age beliefs. It's no wonder that so many aging musicians keep tapping their toes and tossing off notes until the very end.

Nina Kraus, who directs the Auditory Neuroscience Laboratory at Northwestern University and plays the electric guitar in her spare time, found that engaging with music isn't just about hearing. Making music engages the brain, attention, and memory, as well as our sensory and cognitive systems. And the resulting brain changes are more profound in those who regularly make music. According to Nina, it can be any instrument, type of music (including singing), or level of expertise. In her studies, she has found the same brain advantages in professional blues musicians from Chicago and amateur harmonica players. She also found that starting to make music for the first time in later life can benefit your brain and hearing. Nina says that older musicians not only hear music better, but their brains are also better at processing many different types of sound.[14] And some of the skills musicians cultivate over their life span, such as the ability to separate specific sounds from a noisy background, can be taught in a six-week computer-based sound-training intervention that significantly improves hearing among older people.[15]

Nina hasn't studied age beliefs or their impact on the hearing brain, but she said it would follow from her findings that positive age beliefs gleaned from the environment could lead to better hearing in later life and probably other improved senses as well. This is especially likely if the beliefs reduce stress and encourage an engagement with music.

Meaningful intergenerational activities, such as making music together, can also reinforce positive age beliefs.[16] My husband plays violin in a community orchestra with people of different generations. My daughters grew up playing in intergenerational chamber music groups, some that included their paternal grandparents, who are both professional musicians. (Although my musical skills stop at banging out a few very simple piano tunes I learned as a child, I enjoy hearing my family practice and perform.) And when younger and older musicians play together, it's natural for musicians to look up toward those in the group who have accrued considerable experience and expertise.

"Never So Richly Gold": The Aging Senses

The cultural narrative prevailing in the Western world is that the very young are receptive and malleable, and that as we go through the years, we toughen, becoming rigid and unfeeling. But as Penelope Lively, the British, Booker Prize–winning author, writes at age eighty-seven:

> I am as alive to the world as I have ever been—alive to everything I see and hear and feel. I revel in the spring sunshine, and the cream and purple hellebore in the garden; I listen to a radio discussion about the ethics of selective abortion, and chip in at points; the sound of a beloved voice on the phone brings a surge of pleasure. I think there is a sea-change, in old age—a metamorphosis of the sensibilities. . . . Spring was never so vibrant; autumn never so richly gold.[17]

Befitting her name, Lively describes a world aflame with sensuality and tactile engagement. She has written more than forty books, four in the last ten years, of which two are memoirs, suggesting that old age for her has been not only a richly sensual but also an introspective and productive chapter.

As Joan Erikson suggests in her book *Wisdom and the Senses: The Way of Creativity* (published at eighty-eight), creativity in old age both feeds off the senses and in turn nurtures them. She was a professional dancer when she met Erik, her husband-to-be, in Vienna, and thought deeply throughout her adult life about art and creativity, and their role in human development:

> How does one best orient oneself throughout the life-cycle span in order to support, maintain and even enhance the possibility of keeping the senses alive and acute? What activities promote the necessary involvement and are universally, time-honored ways of enriching life? The answer is, of course, that creative activities in general, and specifically all art-oriented making and doing throughout life, offer this fulfillment.[18]

Building on these observations by Penelope Lively and Joan Erikson, I believe a virtuous cycle can exist between positive age beliefs, creativity, and sensory experience in later life. I'd like you to meet Nancy Riege to see what I mean.

Finding the Center of the Labyrinth

Nancy Riege is a sixty-three-year-old artist who lives in the mythically named Northeast Kingdom of Vermont. When we spoke, it was February. The entire northeastern United States was buried under a thick blanket of snow. But Nancy was happy as a kid home from school on a snow day. Winter is her favorite season: she finds the quiet contemplative, and the cold invigorating. She has been making labyrinths—circular walking paths that allow for meditation—over the past thirty years.

They fill a need, she says, that isn't often met in most present-day communities: a path that allows walkers to quiet down, listen to their stillness, and search for a center. In the winter, she carves the labyrinths out of snow, "two snowshoes wide," she explains, with plenty of space for whoever chooses to walk them. In the summer, she mows them into the grass and sprinkles the paths with turkey feathers.

When we spoke, Nancy had constructed five labyrinths for the people of Greensboro, the most that she's ever created at once. Her latest one stands in front of the local school. From their classroom windows, the village's fifth and sixth graders watched her build it over the course of a week and then enjoyed strolling in it with people of different ages the following week. (We will return to the paths of Greensboro in the Afterword.)

For millennia, labyrinths have adorned coins, petroglyphs, fields, pots, and baskets in places as disparate as Java, Australia, and Nepal. In many traditions there is a strong link between labyrinths and ancestors—the labyrinth can serve as a symbolic path to the ancestral home or the ancestors themselves.[19]

As Nancy described her work, I was reminded of Tibetan monks who create sand mandalas by carefully dropping colored grains of sand into intricate geometric patterns. When the pattern is finally finished, which can take up to three years, the monks destroy it, as a reminder of the impermanence of life. Nancy is drawn to making labyrinths for the same reason: they generally last only a few weeks, at most a season, and keep her rooted in the present. She knows some people make labyrinths out of heavy stones but loves that the ones she plows out of snow can be blown over with the next snowdrift.

Nancy is almost mystically drawn to the idea of the center. The labyrinth's center is the first thing she decides on, "by walking around the area. I just close my eyes and sort of feel in my body where the center is." Then she looks around to get "the feeling of the land, the topography, the grade and wind, the nearby sounds, the outer limits." Then she walks around some more, closing her eyes from time to time, to choose the direction of the path, the size and the turns. She describes her creative process as being very intuitive.

Halfway through our conversation, I ask Nancy about her childhood, and soon she's telling me about her grandparents. "Those qualities I gravitate to, those things I need—balance, stillness, just *being*—my grandparents had those in spades." She loved her parents but often found herself wishing they would slow down and "step off the hamster wheel." She concedes "they had to earn a living," but she found their approach to life so different from that of her grandparents, who lived nearby. Although they were busy, too (they ran the volunteer organization for a local orphanage), "they weren't running around trying to do so much, they were comfortable just being." So it's with pleasure that she talks about becoming more like them with aging.

As she gets older, Nancy has become more focused on symmetry and balance (important considerations for labyrinths, which often consist of paths meandering through nestled circles). And her labyrinths, she's found, have gotten better. She wondered aloud if it had something to do with the kind of person she was becoming—someone more like her grandparents, who looked for ways to contribute to their community and were eager to spend time outside "just existing, only more vividly."

Soon after we first spoke, Nancy sent me photographs of her latest labyrinths with a follow-up message explaining that she values elders because they "can overcome the boundaries, walls, or 'rules' we learn to walk within as we navigate the world. With elders, I see the hope of unlearning those unspoken rules that I believe have kept us separate from our true selves—of our true presence." She ended her note with a quote from the elder Benedictine monk David Steindl-Rast: "May you grow still enough to hear the stir of a single snowflake in the air, so that your inner silence may turn into hushed expectation."

"Alterstil" and the Deep Well of Lived Experience

When sixty-eight-year-old Henry Longfellow was asked to speak at his fiftieth class reunion at Bowdoin College, he read a poem he'd written for the occasion:

> It is too late! Ah, nothing is too late
> Till the tired heart shall cease to palpitate. . . .
> Chaucer, at Woodstock with the nightingales,
> At sixty wrote the Canterbury Tales;
> Goethe at Weimar, toiling to the last,
> Completed Faust when eighty years were past. . . .
> What then? Shall we sit idly down and say
> The night hath come; it is no longer day? . . .
> Something remains for us to do or dare;
> Even the oldest tree some fruit may bear; . . .
> For age is opportunity no less
> Than youth itself, though in another dress,
> And as the evening twilight fades away
> The sky is filled with stars, invisible by day.

Even though this poem was written 150 years ago, it feels contemporary in its claims and concerns. Longfellow gently but firmly disputes the thought that old age is a time when opportunities are lost. Rather, he contends they can become recognizable for the first time, in a new form.

Dean Simonton, a California psychologist who studies creativity in later life, found when he investigated what he calls "creatives" across time and cultures that "the ratio of bull's eye to the total number of shots stays the same with age."[20] In other words, the quality of creative work remains constant across our life spans. In addition, there are many examples of what he calls "late bloomers,"[21] or "creatives" who peak in later life. This somewhat depends on the field one chooses, since certain areas, such as theoretical physics and pure math, tend to produce early peaks, whereas fields that build on accumulated knowledge, like history and philosophy, tend to produce later peaks. The philosopher Immanuel Kant, for instance, wrote many of his most important works in his late fifties and sixties.

One distinct advantage of age is experience. The violinist Arnold Steinhardt noticed that as he and his fellow performers in the Guarneri Quartet grew older, they became more sensitive to composers' emotions, which is consistent with what scientists are learning: we become better at reading others' feelings as we age.[22] Over the course of their careers, the quartet played one particularly dramatic and haunting piece hundreds of times, Schubert's *Death and the Maiden*, written when the composer was dying in 1824. When Steinhardt listened to two recordings, made twenty years apart, he noticed that the musicians' approach had markedly improved. Over time, the quartet began to slow down the tempo in the third and final movements to better capture the intentions of the dying composer. "Is it encouraging," Steinhardt wondered, "that after so many performances we are still improving, or is it depressing that we have played so long without hearing something so obvious as the right tempo? Perhaps today's tempo is possible only because it is an outgrowth of every single one that preceded it."[23]

Art historians and researchers of creativity have found evidence for an old-age style, or "Alterstil," which includes "drastic changes in technique, affective tone, and subject matter."[24] They see it marked by an increased sense of drama, a more instinctual technical approach, an expansion of perspective, and a reliance on intuition and the unconscious. As the artist Ben Shahn observed at sixty-six, there is a growing awareness of the inner life of the creative process. He drew more and more "from the most remote and inward recesses of consciousness; for it is here that we are unique and sovereign and most wholly aware."[25]

Michelangelo's two *Pietàs* that he sculpted fifty years apart help illustrate this "old-age style" and the increase in creativity that often comes with age. At the age of twenty-three, he sculpted a biblical scene, now displayed at the entrance of St. Peter's Basilica, depicting a youthful Mary holding her dead son, Jesus, in her lap and draped across her arms.

As a seasoned artist, Michelangelo felt empowered by his age beliefs. He is known for making the statement in later life, *Ancora imparo*, "I'm still learning."[26] At the age of seventy-two, he sculpted the same scene, but in an unconventional way. The *Florentine Pietà* depicts three intertwined figures: Jesus, Mary, and Mary Magdalene, in the lower two-thirds, and an old man in the upper third. The old man, who stands behind and supports the other three, was a self-portrait. The artist intended this sculpture to decorate his own tomb. In the first *Pietà*, Mary gazes down at Jesus without any sorrow marring her face. In the second, she looks distraught as she holds him up; she can't do it alone. The old man is helping her. Nor is she alone in her suffering—their figures are physically and emotionally intertwined. It is a more tender, more human rendering of love and grief.

Joseph Turner, the nineteenth-century English landscape and seascape painter, known for his dramatic depictions of ocean and light, also exemplified this expansion of perspective in his later years. According to his biographer, "Turner's vision grew broader and less specific" as he entered his sixties, fueled by a "growing disdain for petty details" and resulting in "the grandeur of Turner's later pictures."[27]

The photographer Jo Spence was in her fifties when she switched her focus from commercial photography, such as wedding photos, to a groundbreaking kind of documentary photography that confronted larger social issues, such as prejudice by the medical profession. In her memoir, *Putting Myself in the Picture*, she describes a hospital experience she had as an older woman.[28] A doctor arrived at her bedside with a gaggle of medical students and looked over her chart. Then, after delivering the verdict of cancer to the students, he silently drew a cross over her left breast to indicate that it should be removed. Spence then used her camera as a way to cope and protest the way that she had been treated, as if she were not human, by the doctor. She took a series of naked self-portraits. These included one with a cross drawn over her breast and a question sprawled on her body: "Property of Jo Spence?"

In a landmark linguistic analysis of the creative output of ten well-known English-language poets, playwrights, and novelists from the last five hundred years (with equal numbers of male and female authors), psychologist James Pennebaker found that cognitive complexity increased as the writers aged. He conducted the study by analyzing the authors' use of language, such as words indicating "metacognition" or thoughts about thoughts, as in the case of "realize." Pennebaker concluded that the associations between language use and aging in most authors "were quite remarkable." As an example of the robustness of the increasing cognitive complexity with age finding, he pointed out that the American poet Edna St. Vincent Millay and the British novelist George Eliot both strongly showed this improvement with age despite the fact that these women differed in "their genres, nationalities, and centuries in which they wrote."[29]

Similarly, the psychologist Carolyn Adams-Price discovered that older writers tend to engage more directly with emotional meaning, whereas younger writers tend to be more literal.[30] She asked a group consisting equally of young and old people to judge the work of a dozen writers, without indicating their ages. Regardless of their own age, the readers judged the writing of the older writers as better written, more meaningful, and showing "more empathetic resonance." Perhaps, Adams-Price concluded, "late life writing may reflect the positive aspects of late life thinking: synthesis, reflection and even wisdom."[31]

Reinventions

Many older artists reinvent themselves in later life, often to resounding success. The concert pianist Arthur Rubinstein changed his approach to music when he found himself unable to move his fingers quite as fast across the keyboard. He compensated by changing his phrasing, so that he slowed down more before dramatic moments and then sped up to hit the high points of musical phrases.[32] The American folk artist Anna Mary Robertson Moses, widely known as "Grandma Moses," worked in embroidery until her late seventies, when she developed arthritis in her fingers and took up painting instead. She painted every single day until her 101st birthday, generating over a thousand paintings in the course of her late and long career.[33] And Henri Matisse turned from painting to the colorful, ebullient paper cutouts he produced with scissors in the last decade of his life, after surgery made it difficult to stand at an easel. He called this reinvention "my second life."[34] Most critics see it as one of his most brilliant artistic chapters.[35]

In his analysis of the lives of seven older creative people who "changed the direction of the century," Howard Gardner describes how they all made a profound switch in later life.[36] For example, Freud shifted from writing about medical case studies to more broadly investigating ideas of culture and civilization. Martha Graham, who reshaped American dance and helped make modern dance more emotionally expressive, retired from dancing at seventy-five. Then, at seventy-nine, she reemerged as the director and choreographer of her dance troupe, leaving an equally profound legacy on the art form.

Artists are often rejuvenated by the approaching end of their careers, a phenomenon sometimes called the "swan song."[37] The American writer Henry Roth experienced great success at twenty-eight with his first novel, *Call It Sleep*, and then, suffering from writer's block, wrote nothing for the next forty-five years, until he furiously produced *six* novels in his seventies and eighties. He felt his writing helped him look both backward and forward, allowing him to process his earlier regrets and his own mortality. Writing in his later years became, as one of his protagonists puts it, in *From Bondage*, "a window into my remaining future . . . my survival and my penance."[38]

As noted in our discussion of mental health, emotional intelligence and the willingness to engage meaningfully in life review both increase as we grow old. This feeds a powerful current that can irrigate our creative impulses as our drive to find or make meaning in later life translates into renewed or improved creative output.

Liz Lerman: Dancer of Genius

When the dancer and choreographer Liz Lerman turned sixty-nine, she was looking for a change of creative scenery. She joked to her husband, "I've got to change my job, change my house, or change my husband." She kept the husband, but switched the other two, moving across the country from Baltimore, Maryland, to Phoenix, Arizona, where she became a dance professor at Arizona State University. Since then she has been thinking about, engaging in, and helping others find their own creativity. Liz went on to clarify that we don't need to make a major life change to activate creativity; we can change or expand our connections to people.

In graduate school, I once took a dance workshop with Liz that was so innovative and energizing that I've kept a photo of her, taken when she was in her thirties and dancing with three older dancers, above my desk ever since. So when she agreed to talk with me about her creative process, I felt like I was talking to an old friend who keeps me company when I write.

Liz, who won a MacArthur "genius award" for "redefining where dance takes place and who can dance," epitomizes the synergy of positive age beliefs and creative activity. She started dancing with seniors soon after graduating from Bennington College, and first choreographed a major piece for older dancers during the period of enormous loss that followed the death of her mother. Funneling her grief into the dance, she envisioned old angels welcoming her mother into heaven. She had older dancers play the angels. "I didn't know it was going to become a thing," she told me. "I just thought I need to do this piece, and I need old people to be in it." Once she started choreographing for seniors, though, she never stopped. She founded a dance group in Washington, D.C., called Dance Exchange, which became world famous for its reliance on personal story, public participation, and intergenerational dancers. In contrast, due to ageist pressures, most professional dancers in Western countries retire by age thirty-five.[39]

Since then, Liz has gathered countless people of all ages to dance—including seniors who have never danced before and people who never stopped. It's not so much teaching them perfect technique that she is after: it's the loose, primal, joyful thing dancing does to people's bodies and self-perception. "Often, it's like, 'Oh, my god. Look what I'm doing!'" Liz says with a joyous shout. And for both kinds of dancers, Liz believes intergenerational dance helps them strengthen their age belief that older people meaningfully contribute to society.

"Older dancers have movements unique to their age," Liz explains. "When a person moves in harmony with an idea or an emotion, with a vocabulary of movement that is inherently personal to their body, the result is something staggeringly beautiful."

Thomas Dwyer is an eighty-five-year-old dancer who has been a member of Liz's company for the last thirty years. A lifelong conservative Republican, retired navy veteran (he was a Morse code operator on ships for most of his career), and six-foot-tall man who describes himself as a "string bean with no muscles," he is the first to admit that he is an unlikely dancer: "When people see me dance, it helps them see that anybody can do it." His first dance workshop took place at the urging of his brother, who'd accidentally signed up for an earlier workshop, thinking that it was an exercise class. Both were soon hooked.

Thomas's favorite dance, "Still Crossing," is about immigration. It opens with old people rolling slowly across the stage, which Liz describes as "the ghosts of my grandfather—all those immigrants present in each of our imaginations," and ends with dancers of all ages, including a dozen seniors, onstage. The first time it was performed was at the Statue of Liberty. In another dance, Thomas does a set of sixty push-ups in his underwear with his feet on a chair. Invariably, people come up to him afterward and tell him they were surprised to see someone his age do that.[40]

In Tokyo, while participating in a residency that brought together younger professional dancers and older Japanese people who were dancing for the first time, Liz noticed that when the older dancers entered the studio, they didn't instantly diminish themselves, the way she's seen countless older Americans do. In the US, "the messages diminishing old people are everywhere. They turn old people inward the way leaves turn inward." She held up a hand and curled it inward, and then curled her body inward, like a leaf shriveling in fast motion. "But once you get them dancing," she said, they bypass all the negative messaging, and a change occurs. She held her hand up again: "Picture the leaf again for a minute. Instead of being brown and brittle, imagine the water moving through it, nourishing it; watch it soften, open up, and move outward."

But it's not just by rethinking their relationship to their bodies that dance transforms older dancers; it's the intergenerational component that alters their sense of potential as older people. One of the reasons intergenerational dance works so well is that people in their sixties and twenties actually have quite a lot in common, Liz says. "In a way, that's because they are at similarly transformative stages in their lives, thinking about the big questions. 'Where am I going? What will I do with the rest of my life?' And those questions hit hard." Younger people are completing high school or college; older people are rounding out a career, or changing their living arrangements. And when they dance together, "they become so close," Liz tells me. Negative stereotypical conceptions of old age break down.

Most people hear "intergenerational and immediately picture kindergartners and grandparents," Liz says, but she thinks there is a particular affinity between younger and older adults. Many young people are searching for sources of love and support, whereas many older people are searching for ways to share these things. Liz has seen countless young people find themselves transformed by the physical company and collaboration of older people, "because they felt loved in a way they were just yearning for." Liz believes that the beauty of intergenerational creative activities is that they give people of different ages the permission and framework to work together in a way that feels welcome.

Now seventy-three, Liz finds herself in the midst of the most generative period of her life. While continuing to teach, she recently designed a free online creativity toolkit ("The Atlas of Creativity"), choreographed a dance about stereotypes of female bodies ("Wicked Bodies"), and is working with the Urban Bush Women, an African American professional dance group, on a project called

Legacy of Change. Her mastery and originality have only deepened, she feels, as a result of her decades of experience. Between teaching and performing and collaborating, she also wants to bring dance to ever-increasing numbers of people who have never danced before. As she grows older, she is constantly thinking about how to help those who come after her: "Legacy isn't just looking back, it's also looking forward."

To go forward, it is important to consider the barriers preventing older persons from becoming as creative and generative as they would like. This is the aim of our next chapter. As author and social critic James Baldwin wrote, "Not everything that is faced can be changed, but nothing can be changed until it is faced."

8

Ageism: The Evil Octopus

The Birth of "Ageism"

Three years before he broke the Watergate scandal that would lead to President Richard Nixon's resignation, twenty-five-year-old reporter Carl Bernstein helped bring to light a different kind of scandal. On a windy March morning in 1969, the young reporter interviewed psychiatrist Robert Butler about the growing hostility of the residents of a Washington, D.C., suburb to the proposed transformation of a nearby apartment building into senior housing. As head of the local Advisory Committee on Aging, Butler had recently met with the neighbors who were wringing their hands in worry that the area would never be the same. The sentiment, Butler told Bernstein, was that "they didn't want to look at people who may be palsied, can't eat well, who may sit on the curb, and clutter up the neighborhood with canes."[1]

Butler, who lived in the same neighborhood, saw this kind of ugly negative stereotyping as no different from the stereotyping of racism or sexism. The same way those two isms negatively label people of color and women, thereby keeping them away from opportunities and power, this prejudice toward the aged, which Butler called "ageism," also deprived older persons of equal rights. It was the first time anyone had given it a name.

In his seminal book *Why Survive? Being Old in America*, Butler later defined ageism as a "process of systematic stereotyping or discrimination against people because they are old." The two components of ageism, he realized, are mutually reinforcing. Negative age stereotypes lead to age discrimination, which then activates and reinforces the stereotypes. In the end, he wrote, "ageism allows the younger generation to see old people as different from themselves; thus, they subtly cease to identify with their elders as human beings."[2]

When I spoke to Bernstein, he remembered his encounter with Butler as a lightbulb moment. He had encountered ageism before in the poor treatment of older relatives, but he had never thought of it as discriminatory or systematic. After his encounter with Butler, though, it was different. As Bernstein recalls, "From a story about a bunch of citizens who didn't want some old people in the neighborhood, it became an article about the phenomenon of ageism. I became affected and more aware of it after that event. It was fear of the other, no different from fear of Jews, or African Americans, or Catholics at one point. Discrimination is discrimination. It's based on fears and stereotypes." Thanks to Robert Butler, the phenomenon of ageism finally came to light. Fifty years later, unfortunately, it continues to prosper.

We've looked at age beliefs' profound health impact. Now, let's look at how negative age beliefs operate on a societal level in silent, complex, and often deadly ways, intertwining and flexing like octopus tentacles.

Ageism: The Silent Epidemic

It's common for people to downplay or whitewash the damage of ageism. Sometimes, when people learn what I do for a profession, they tell me that ageism is not a serious issue, or that it doesn't exist.

Or even that ageism is the fault of the old, as in a comment I recently heard: "Ageism is just a mirror held up to older people falling apart." The trivialization of ageism and blaming older people for the prejudice and discrimination they face compounds the problem.

When I ask audience members during presentations how many have directly experienced or observed someone else experiencing ageism, a majority of the hands go up. Today, 82 percent of older Americans report encountering ageism regularly,[3] and I have found examples of ageism in every country I have studied.

How can it be that ageism remains a nonissue for some, while so many have experienced it? A recent World Health Organization report concludes that "people don't recognize the existence of institutional ageism because the rules, norms and practices of the institution are long standing, have been ritualized, and are seen as 'normal.'"[4]

One of the most insidious ways ageism works is by ignoring older people. Just look at their near absence from movies, advertising, and TV shows, from national conversations on urgent public-policy issues, in research trials, and in so many other realms of contemporary life.

This neglect is most striking in times of crisis, when older people typically come last. After Hurricane Katrina, animal activists evacuated dogs and cats within twenty-four hours, whereas many older people were abandoned in their homes and left to face the rising flood until medical teams finally arrived to rescue them, sometimes up to seven days later.[5] In the early stages of the COVID-19 pandemic, 40 percent of US deaths were of older persons in nursing homes (even though less than 1 percent of the population lives in nursing homes), where local governments and nursing-home administrators failed to provide adequate protective equipment, COVID tests, or quarantine locations, all of which were basic resources offered to lower-risk younger people at most universities and colleges.[6]

My Experience as a Target of Prejudice

Many prejudices rely on invisibility. Bigots often deny that they are racist or refuse to accept that racism is alive and well. Sexists often argue that women no longer face prejudice. I experienced this kind of denial with anti-Semitism.

There weren't a whole lot of Jews living in Vancouver when my parents moved our family there from Boston to take positions at the University of British Columbia. As I was the only Jewish kid in my new second-grade class, the teacher asked me to stand in front of the group to justify why my family didn't celebrate Christmas. One classmate told me I couldn't play with her because I was Jewish. A group of boys threw pennies on the ground and told me to pick them up.

When I told my mother about these events, she consoled me and brought it up with my teacher and classmates' parents, only to be told that this wasn't really anti-Semitism, just a cultural misunderstanding. This was the first time I encountered this strange dimension of prejudice, which is that people will try to pass it off as something lesser than it is.

In childhood, I had a nightmare of being chased through dark woods by Nazis led by barking Doberman pinschers. My brushes with anti-Semitism pale with the *actual* nightmares that my great-grandparents and grandparents survived in Europe, but I am only two generations removed from horrors whose details still visit my dreams. My paternal great-grandfather was one of just a handful of survivors when his Jewish town in Lithuania was burned down by Cossacks. By hiding in a cabinet, at the age of ten, my maternal grandmother narrowly escaped Russian soldiers carrying out a massacre of Jews.

Anti-Semitism was a personal experience for me, but my challenges at school and my relatives' suffering in Europe were also rooted in structural prejudice carried out by those in power. Those early experiences made me sensitive to, as well as curious about, the causes and expressions of prejudice. When I first encountered ageism in an institutionalized context, in my first job at the geriatric unit of a hospital, I felt as though I had been given an opportunity to fight prejudice in a way I hadn't been able to as a child.

The Age-Stereotype Paradox

Ageism has a way of flying in the face of reality. Let me offer you a thought experiment to illustrate what I mean:

Go back in time two hundred years, all the way to the 1820s. Photography has just been invented, and tracks are being laid down around the world to accommodate a new contraption, the steam engine locomotive. Now, from this sepia-tinted vantage point, gaze into the future and guess whether age beliefs will improve, stay the same, or become more negative. To give you a bit of an edge, I'll throw in some hints about a few trends that will take place in the next two hundred years from then: Older people will live longer, and their overall health will greatly improve. They will make up a larger percentage of the population, which means there will be more opportunity for intergenerational engagement. A series of laws will be passed that ban age discrimination. And to top things off, people will become much more positive in their attitudes toward other previously marginalized groups.

Now, what do you think? Over the next two hundred years, will age beliefs improve, stay the same, or become more negative?

Most people assume that age beliefs become more positive. This is what I would have concluded, based on these trends. What has actually happened, though, is the opposite: in the US, views of older people started as positive and have become more negative in a steady, linear way.[7] My team discovered this when we developed a computer-based linguistic method to systematically examine age belief trends over two hundred years. (Previous systematic analyses had been limited to no more than twenty years.) To conduct our analyses, we examined four hundred million words from printed texts with a newly available database called the Corpus of Historical American English. Where is the flood of negative age stereotypes coming from? And why, in the face of all our progress, has it not dried to a trickle?

Causes of Ageism: Lizard Brains and Corporate Greed

The stubborn persistence of negative age beliefs has both individual and structural causes. Although they are distinct, both are deep-rooted. On the individual level, there are numerous psychological processes that make it easy to express ageism without thinking about it. Structurally, ageism is embedded in institutions and people with power.

Ageism in individuals starts with the fact that we assimilate age stereotypes early in childhood, long before they become self-relevant. At this stage, we accept them without any resistance. And given that these stereotypes are often presented by people with authority whom we trust (teachers, authors, parents), it is easy to accept them as truth, which then becomes the blueprint for how we conceptualize old people throughout the life span.[8]

Negative age beliefs fill a societally created psychological need by those who are not old to distance themselves from those who are old in ageist cultures. This distancing takes a physical form when spaces frequented by older persons are avoided and a psychological form when they are dehumanized through stereotypes. Some young people feel a need to distance themselves from older persons because they appear as feared future selves.[9] The process entails a vicious cycle when the negative age beliefs, with their prevailing sense of debilitation, present a grim picture of old age, which, in turn, intensifies the effort to create a sense of distance, which can then reinforce their negative age beliefs.

Yet another individual-level cause of ageism is that it often operates without our awareness. So even if people think of themselves as fair-minded, they may in fact be engaging in ageism.

Compounding this problem, negative age beliefs are often accepted and expressed, even when they go against experience. For instance, a person might make a joke about an aging person showing senility or incompetence, even though it's known they're as sharp as ever.

The main structural motive for ageism is that it is often quite profitable, both financially and as a means of preserving power. A former professor of mine, the anthropologist Robert LeVine, said that a good question to start with when investigating a cultural phenomenon is: "Who profits from the status quo?"

A number of commercial enterprises make a stunning profit from promoting negative age beliefs. These include the antiaging industry, social media, advertising agencies, and companies that are based on creating fear of aging and an image of older persons as inevitably declining. Together, these sectors generate over a trillion dollars a year and have been steadily growing, largely without regulation.[10]

When Ageism Begins: Cartoons and Fairy Tales

Looking for something to watch on a recent flight, I saw that my options were limited to movies I'd seen or to ones that seemed like they weren't right for the occasion (the trailers showed things like planes blowing up). So I settled for a recent Disney children's movie, a remake of the classic story "Rapunzel," called *Tangled*. In this version, to benefit from the magical antiaging properties of Rapunzel's hair, the witch, who is portrayed as an old woman, locks Rapunzel in a tower. At the end of the film, after losing access to Rapunzel's hair and eternal youth, the witch instantly ages. She shrivels into a gnarly crouch, her hair turns from black to gray, and she develops sunken eyes and bony hands. "What have you done?" she screams at Rapunzel's rescuer. And then, she dies. Yet there was no antiaging theme in the original Brothers Grimm story—Disney gratuitously added it. Perhaps Disney added the age stereotypes to the movie to increase ticket sales to families with children by uniting the audience with a narrative that denigrates an older person and suggests that old age should be feared and avoided.

One of the first songs I learned in nursery school had a chorus you might be familiar with: "I know an old lady who swallowed a fly. I don't know why she swallowed a fly. Perhaps she'll die!" As the song progresses, she goes on to swallow increasingly large insects and animals, including a dog and a horse. When I first learned this ballad about the strange old lady, my classmates and I thought it was hilarious.

As children, we often first encounter older people in songs, nursery rhymes, and stories. In many Western countries, these older characters are often villains or objects of pity or ridicule.[11] As a result, it is not surprising that children in these countries fear becoming old.[12] Shown drawings of a man's face at four stages of life, 80 percent of schoolchildren said they would rather spend time with the younger version of the man, and when asked what kinds of activities they would like to join in with the oldest version of the man, an example of a response was: "Bury him."[13] Children as young as three shrink from older people and display beliefs that are unambiguously ageist.[14] The age beliefs we take in as children form the basis of our later-life age beliefs.[15] The age code is accretive, built on successive layers, the foundations of which are laid down in childhood. And because these beliefs aren't yet self-relevant, children see no reason to resist them, especially when they are encouraged by those that they admire.

A friend told me that the principal of her children's elementary school recently sent home a notice announcing that he was holding a "Dress Like a One-Hundred-Year-Old Day" to celebrate the one hundredth day of class. He encouraged parents to send their kids to school in white wigs, large plastic glasses, miniature toy canes, and walkers (all available at the local party store). This is a popular school practice, it turns out. Websites abound with advice for parents on how to dress their children in drab colors and teach them to "hobble around" with canes and to "go a little overboard as a stereotypical old person to make it more fun!"[16]

This new childhood tradition could contribute to a long-term risk. Using the Baltimore Longitudinal Study of Aging, as can be seen in Figure 5, we found that younger people who assimilated more-negative age stereotypes were twice as likely to have a heart

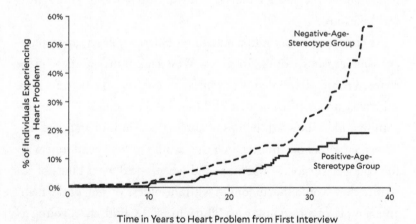

Figure 5: Younger Persons' Negative Age Beliefs Increase Risk of Experiencing a Cardiovascular Event After Age 60.

attack or other cardiovascular event after turning sixty than those who assimilated more-positive age stereotypes.[17] What we teach our children about aging matters, not just for how they treat others, but for their own health.

Teenage Botox and the Antiaging Industry

The rapidly growing global antiaging industry generates half a trillion dollars a year by selling pills, creams, tinctures, elixirs, hormonal supplements, "testosterone boosters," and procedures that falsely claim to halt or even reverse aging.[18] The industry profits by promoting images of aging as something to fear and avoid.

Not long ago, I was eating lunch with a colleague when she raised her eyebrows after receiving a text from her seventeen-year-old daughter. She read me the text: "OMG, just got my first wrinkle. For my birthday can I get preventative Botox?"

It's easier to sell Botox and face creams once you have ensured that your prospective customers are frightened of any physical sign of aging. Recently, while I was waiting to see a doctor for a routine visit, the large screen in the waiting room streamed ads for various antiaging procedures, including Botox. "Hey, wrinkle face!" yelled comedian and talk-show host Ellen DeGeneres in one commercial. She went on to explain that an intervention is needed to avoid looking like a prune. Although she likes eating them, she adds, "I don't want to look like one!" The mass marketing of antiwrinkle products and procedures has become a common practice.[19]

That's how you get seventeen-year-olds to become afraid of wrinkles, a natural and universal occurrence—by telegraphing the message that you can and should avoid aging, that you can't be beautiful *and* wrinkled, both valued and aged. It's no surprise, then, that in the last two decades, Botox injections have more than tripled among young adults.[20] This has helped the branch of the antiaging industry that targets wrinkles to generate profits of nearly two hundred billion dollars.[21]

Botox shots to prevent future wrinkles are now routine among American women in their twenties and thirties. A recent *New York Times* article explains that "baby Botox" injections for young adults are becoming "destigmatized," but it glosses over the practice's roots in age stigma.[22]

Lest you think wrinkles and the antiaging industry's profits from fear of aging are predominantly female issues, there has also been an increase in ads vilifying receding hairlines; the rate of surgical hair transplants among men has shot up 60 percent in the last five years. Societally, we now balk at virtually any sign of aging.[23]

One typical antiaging advertisement declares its product is the "Good Housekeeping Anti-Aging Gold Winner" of the year and goes on to describe the ways in which it can "help with the battles against the 5 signs of aging hair."[24] A review of nearly one hundred websites that sell antiaging products revealed a prevailing approach: We are at war with old age, and consumers who don't purchase antiaging products have let themselves go and given up the fight.[25] These sites almost always offer ineffectual or even harmful "cures" at great expense. Human-growth hormone, for instance, is the ultimate (and pricey) elixir, advertised for those who "don't want to age rapidly but would rather stay young, beautiful and healthy all the time."[26] Meanwhile, this hormone that is sold in many local stores can also increase your risk of diabetes and cancer.[27] To maximize profits, lobbyists for the antiaging industry have helped set up loopholes that shield some of the industry's products and marketing from federal regulations.[28]

The antiaging industry warps our conceptions of beauty by stigmatizing not just the process of aging, but older people themselves. As the geriatrician Tom Perls put it: "The hucksters' sensationalized images of older people as withering and frail individuals staring at nursing home walls reinforce our youth-oriented society's inaccurate and bias-engendering perceptions of aging. Anti-aging has become synonymous with anti-old people."[29]

The Not-So-Silver Screen: Ageism in Pop Culture

Back when I was in grad school in Massachusetts, I joined a local chapter of the Gray Panthers, an intergenerational antiageism activist group, where I was part of the "Media Watch" team. We hunted for examples of ageism in newspapers, magazines, movies, and on the radio.

One month, we focused on news articles that trafficked in the popular, fear-provoking trope that older people are what some columnists called "greedy geezers."[30] The examples were endless. But in a direct rebuttal of this idea, a study I later conducted showed that, in fact, older people are actually more likely to oppose programs aimed at benefiting their age group (Social Security, Meals on Wheels, and Medicare) than younger people.[31] I also learned that older people are more likely to volunteer and give money to nonprofits than younger people, and they provide the equivalent of billions of dollars in unpaid caregiving to family members and friends.[32]

Television is another vector for the spread of ageism. Older people watch more TV than any other age group, but only 2.8 percent of characters on TV are old, and usually they're relegated to minor and unfavorable roles.[33] This lack of meaningful roles is likely rooted in the fact that networks and studios often exclude older writers and rely on advertisers who prioritize what they call the "key demographic" (aged eighteen to forty-nine), out of the misguided belief that only those below fifty will buy new products.[34]

While the TV and movie industries have improved their inclusion of gay and female characters in leading roles,[35] older characters continue to get short shrift. When elderly characters *are* depicted in Hollywood films, it is often as exemplars of cognitive and physical decline (*Iris* and *The Father*), grumpiness (*Grumpy Old Men* and its sequel, *Grumpier Old Men*), and horror—*The Visit* depicts grandparents who plan to murder their grandchildren and *Old* shows a vacation falling apart and striking terror when a family starts to rapidly age. There are a few exceptions that depict aging with complexity and verve, such as the American TV show *Grace and Frankie* and the British TV show *Last Tango in Halifax*, but by and large, older characters are still marginalized. In 2016, characters aged sixty and older represented just 11 percent of all speaking characters across the one hundred top-grossing films, and of those films in which older people actually managed to make it onto the screen, 44 percent included ageist comments.[36]

As of yet, there's no widespread acknowledgment of how ageist Hollywood is. The leadership of the Academy of Motion Picture Arts and Sciences, the body that votes on the Academy Awards, recently mandated that movies must include actors from underrepresented racial groups or other marginalized backgrounds (women, LGBTQ, or disabled actors) to qualify for an Oscar.[37] Announcing the move, the Academy's president said it was time for movies "to reflect our diverse global population."[38] But there was no mention of including older actors.

Twenty years ago, the two-time Academy Award–winning actress Geena Davis started an institute to work toward gender parity on screen.[39] She says, "It wasn't until I had my daughter that I saw that there was profound gender inequality in the movies and TV that are made for kids. Surely in the 21st century, we should be showing kids that boys and girls share the sandbox equally." After fifteen years of offering fellowships to female writers and directors and raising awareness about the exclusion of women from writing and directing roles, she felt she had helped achieve some of what she set out to do. On-screen, at least, there is something close to equal representation of male and female characters today, though not yet when it comes to writing and directing.[40]

Now, Davis has turned her attention to exposing ageism. "Once I had a four in front of my age, I fell off the cliff. Suddenly, the great roles were incredibly scarce. It was a big difference."[41] In its 2019 survey of the thirty top-grossing films in the US, UK, France, and Germany, the Geena Davis Institute found that not a single woman over fifty had been cast in a leading role. "I knew it was bad," she says, "but this really drove home how very dismal it is."[42]

Our TV and movie culture warps our understanding of what old age and older people are really like. My team found that those who watch more television over their lifetimes have more-negative age stereotypes.[43] And when members of marginalized groups don't see themselves represented on television, or in books or ads or online or other media, it can lead to lower self-worth.[44]

The exclusion of individuals based on their older age extends to the fashion world, where most models are closer in age to childhood than middle age. During a recent Fashion Week, when clothing designers take over New York City to display their latest collections, a trio of *New York Times* journalists uncovered some of the problems plaguing the industry.[45] Their article featured twelve models; ten of them were in their twenties, and none over thirty-two. Even these young models had experienced ageism. Renee Peters, a twenty-eight-year-old who was recruited in a Nashville mall by a modeling agent when she was just fourteen, laments, "I was at a casting call yesterday. I looked around and felt like everyone must have been sixteen, seventeen, eighteen. And here I am past twenty-five and really questioning, am I still beautiful? Do I still hold worth?"[46]

Ageism as Clickbait

Twitter, Facebook, YouTube, and Instagram now make up one of the largest profit-making sectors in history. Facebook alone has nearly two billion users—almost a third of the world's population. Attention is the currency most coveted on social media, an industry that gave birth to the so-called attention economy, and outrageous, opinionated, prejudicial content ("clickbait") attracts the most attention.[47] The more-negative clickbait is toward marginalized groups, the more effective it will be (the more clicks it receives), so that advertisers will spend more buying ads.[48]

To see how age beliefs played out on Facebook, my team analyzed all publicly available groups having to do with older people and found that 74 percent of these groups vilified older people, 27 percent infantilized them, and 37 percent advocated banning them from public activities, like driving and shopping.[49] One British group advocated banning older people from stores on the grounds that "they smell of wee, they park badly in the carpark resulting in a [sic] wasted and exceptionally valuable parking spaces. Either age-check them at the door or implement a voluntary euthanasia programme. I'd happily volunteer to top one of them." (*Top* means "kill" in British slang.) We reported the ten most offensive sites to Facebook as examples of hate speech; a year later, the sites were still around.

At the time of that study, hate speech on the basis of sexual orientation, gender, race, and religion was prohibited on Facebook, but ageism was left out. In an updated version of Facebook's community standards, older people are now protected against hate speech, but only if they are tied to *another* protected group, so that inveighing against "old women," for instance, is banned, but not if it's against old people in general, as in the British example above.[50] On its own, ageism is condoned, unlike other kinds of prejudice.

In the first three months of the COVID pandemic, on Twitter, 1.4 million people in over a hundred countries liked or shared tweets containing the term "boomer remover," which mocks the idea of older persons dying from COVID.[51] In another study, psychologist Karen Hooker and her team, using a sophisticated computer method, found that 33 percent of tweets that mentioned Alzheimer's disease did so in a way that mocked older people.[52]

Social media sites are the perfect medium for users to spread ageist stereotypes. The sense of anonymity they afford removes fear of consequences and encourages extreme, provocative, and hateful speech.

It's not just ageist hate speech that spews unchecked on social media; blatant (and illegal) age discrimination runs rampant as well. Social media companies pull together all kinds of data about their users and utilize that data to figure out who is viewing the ads. Based on this data, social media sites know who are older users and exclude them from certain housing ads, credit offers, and job listings—effectively slamming a door in the face of older people.[53]

A housing watchdog group found that Facebook was allowing ads to exclude older potential renters in the same D.C. region where Robert Butler first identified ageism fifty years ago. Ageism endures, only in a more covert and structural way. According to these fair-housing activists, this D.C.-area age discrimination is "not being carried out by one or two small players who inadvertently misused digital tools, but involved several leaders that manage hundreds of thousands of apartments nationally. They paid substantial sums to Facebook to categorically withhold their ads from older people."[54]

In 2019 alone, Facebook settled five lawsuits related to age discrimination; yet, the practice of digital age discrimination in housing and hiring continues.[55] The companies paying for online ads for job openings that purposely exclude older people, by posting the ads in a way that they can not see them, are ones you are familiar with:

Target, UPS, State Farm, Amazon, and Facebook itself, which has a median employee age of twenty-eight.[56]

Agelining: Ageist Spaces

Though the percentage of older individuals in the larger US population has steadily grown over the last one hundred years, intergenerational contact has seen a steady decline. In the course of that period, the US has gone from being one of the most age-integrated to one of the most age-segregated countries in the world.[57] Households are becoming less age-diverse. In 1850, 70 percent of older Americans lived with their adult children and 11 percent lived with a spouse or alone; by 1990, only 16 percent of older individuals lived with their adult children and 70 percent lived with a spouse or alone.[58] In fact, even with redlining and rampant racial segregation, our neighborhoods are now equally segregated by age and by race.[59] The problem is not limited to the US. In 1991, a British child had a 15 percent chance of living near a person over the age of sixty-five; today that has dropped to 5 percent.[60]

One of the factors behind age segregation is the enduring and misguided societal perception that it's somehow beneficial or natural to keep young people away from older ones. But other types of segregation, such as race-based or gender-based, are considered harmful by policymakers, scholars, and the general public.[61] In New Haven, where I work, city planners have built senior housing in locations that are isolated by highways or waterways, almost as if to quarantine the population. This kind of physical segregation drops the likelihood of everyday casual interactions between the young and old, on street corners or in parks, to basically zero.

This lack of contact is an incalculable loss for both the younger and older members of the community. It not only weakens empathy and social bonds between young and old, but removes the opportunity to counter younger people's negative stereotypes about older people.

Ageism Is a Full-Time Job

Consider this scenario: You've been working as the head of marketing for a large company and have been praised over the years for your many creative strategic initiatives. One day, however, you make a decision that others in your company support, but that your boss disagrees with. You explain your reasoning. In response, your boss tells you that you should quit. He doesn't exactly fire you, but says your thinking is "outdated" and suggests that maybe it's "time to retire."

This is what happened to sixty-one-year-old Gray Hollett, who was left devastated by this experience with ageism.[62] And Hollett is far from alone. Search "ageism in the workplace" online and you'll find hundreds of stories like his: people nudged out of the company based on being older.[63] Two-thirds of workers in America said they have witnessed or personally experienced age discrimination in their place of work and, of those, 92 percent said it was a common occurrence, according to an AARP survey.[64]

In the US, discriminating against someone in a workspace on the basis of their age is illegal. The Age Discrimination in Employment Act—ADEA—passed in 1967 says so. But this law exists in name more than deed.[65] The ADEA doesn't allow for the recovery of compensatory or punitive damages, which means lawyers aren't incentivized to take on age-discrimination-related cases and it can be very expensive for older people who have been mistreated to file these lawsuits. Not only that, but the ADEA only applies to people who already have jobs—not job seekers.[66] So if you're turned away from a job because you're "too old," the ADEA won't help you.

The irony behind all this is that job experience, which comes with age, is what most often contributes to success in later-life employment.[67] When Ben Duggar, a botanist specializing in soil health, was forced to retire from his teaching post at the University of Wisconsin at age seventy, he was subsequently hired by Lederle Laboratories, where, at age seventy-three, he isolated a compound called tetracycline, which has become the world's most widely prescribed antibiotic.[68]

Older workers are not only capable of remarkable breakthroughs; they're more dependable and have less turnover, less absenteeism, as well as fewer accidents.[69] Yet, ageism runs rampant at every stage of the employment cycle. When my team looked at ageism in the workplace (both white-collar and blue-collar) in forty-five countries, we found that older workers were significantly less likely to be hired than younger job applicants and, when hired, were significantly less likely to be trained and promoted.[70]

A recent Harvard Business School study at a BMW plant in Germany demonstrated the benefits of retaining older workers. It found that an age-integrated assembly line increased productivity, reduced absenteeism, and led to fewer car defects. The cherry on top? At the end of the study, none of the workers wanted to leave the age-integrated team.[71] Chip Conley, who at the age of fifty-five helped set up similar age-integrated teams at Airbnb, found that these teams are successful because "older workers know how to frame problems and create accountability for results."[72]

A Sickness in Health Care

The medical field is supposed to help and heal, but it doesn't always do that. I have no bias against physicians (my husband is a wonderful doctor), and vaccines and procedures have likely saved my life and the lives of family members, as they probably have yours.

But too often, the medical and scientific approach to cognitive and physical aging is to frame it as the gradual deterioration of various biological characteristics, rather than as a time that can include positive changes resulting from a range of factors including experience.[73]

One reason Western medicine so heavily relies on negative age stereotypes, with their narrative of inevitable decline, is that it's profitable. The multibillion-dollar "medical disability complex," as Carol Estes calls it, is based on expensive procedures, devices, and pharmaceutical drugs, which are more profitable than prevention efforts, such as exercising, or the challenging but necessary task of trying to tackle the societal causes that often contribute to disability and disease in the first place.[74]

When aging is seen as a strictly biomedical phenomenon and the social determinants, such as ageism, that play a crucial role are ignored, doctors are apt to dismiss treatable conditions as standard features of old age (for instance, back pain or depression).[75] The more doctors confound aging with illness, the more it reinforces the view of aging as a pathology, which can lead to the undertreatment of elderly patients. For if doctors expect the health of their older patients to decline, the doctors are less likely to try to help them improve.

Imagine waking up with shooting back pain one morning and finding it hard to walk, and going to the doctor only to be told, "What do you expect? You're old." This is exactly what happened to one participant in a study led by geriatrician Cary Reid that looked at why older adults don't always seek or receive care for back pain.[76]

Many doctors know less than they should about the normal health of older people, and what they *do* know is often influenced by negative age stereotypes. For example, 35 percent of doctors think it is normal for older people to have high blood pressure (it is not),[77] and many fail to gather sexual histories from their older patients, even though individuals over the age of sixty-five are the fastest-growing age group contracting HIV and AIDS.[78] This sets up these doctors to miss diagnoses of STIs, erectile dysfunction, or decreased libido.

Where do many doctors develop their negative and often false perceptions of older patients? Unfortunately, they're often acquired in medical school. The first time that medical students encounter an older "patient" is often in the form of an older cadaver destined for anatomical dissection.[79] All medical schools require training in pediatrics, but few require training in geriatrics, partly because there are too few geriatric specialists to teach them—a vicious cycle.[80] A study found that as medical students progress through their training, their views of older patients become more negative.[81]

When Robert Butler was in training, he discovered that older hospital patients were referred to as "GOMERs"—Get Out of My Emergency Room—a phrase that continues to be used to this day. The National Director for Patients and the Public in the British Department of Health described how medical staff "often dehumanize older patients by calling them 'crinklies,' 'crumblies' or 'bed blockers.'"[82] Butler explains that in his case he "was shocked by the medical lexicon concerning older persons, abounding as it did with cruel and pejorative terms."[83] That's when he decided to go into geriatrics. The negative age beliefs he encountered in medical school were completely at odds with the associations he had of the vital and powerful grandmother who raised him.

Some medical schools train future doctors to learn about older patients by equipping their students with glasses that blur their vision, leg weights to restrict their movement, and headsets that block hearing. To complete "the aging experience," as the training is called, the students are then sent to different "activity stations."[84] In one of the stations, the students experience social isolation at a simulated dinner party where they're left out of conversations. Although the goal of this training is to instill empathy, the effect may be the reinforcement of negative stereotypes if medical students are led to assume that their future older patients will be frail and deficient, rather than thriving, functional people who can enliven a dinner party.[85]

Signaling that older patients are less valued, the health-care system pays geriatricians less than specialists in many other medical fields.[86] Not surprisingly, the field of geriatrics is experiencing an extreme shortage of qualified doctors in many countries. Meanwhile, a survey found that geriatricians tend to enjoy their work more than other specialists because of the gratification they receive from interacting with older patients.[87]

Negative age stereotypes account for why many doctors are less patient, less engaged, and less likely to explain the details of conditions or treatments to older patients, which often leaves them without the information they need to care for themselves when recovering from illness.[88] These stereotypes also lead to the undertreatment of older patients.

In a systematic review of studies investigating how ageism impacts older persons' health, my team found that in 85 percent of studies on health-care access, providers discouraged or outright denied older patients access to certain treatments, compared to younger patients who were identical in every way except for age. In all forty-five countries included, ageism worsened health outcomes for older people.[89]

In spite of this, ageism in health care is not yet seen as a widespread public health or human rights issue. To help policymakers visualize its impact, I teamed up with an economist and statistician to put a price tag on the health costs resulting from ageism.[90] We found that it totals $63 billion per year in the US,[91] which is more than the cost of morbid obesity, one of the most expensive chronic conditions in America.[92] That's how much we stand to save in preventable health-care costs, and it's a conservative estimate, since we only considered eight health conditions and didn't include the cost of lost wages.

"Of all the forms of inequality," Martin Luther King Jr. stated, "injustice in healthcare is the most shocking and inhumane."[93]

Intersectionalities and Ageism

I have described some of the many domains in which ageism operates, but the tentacles of ageism don't always neatly latch on to the separate spheres of our lives. These spheres overlap; the tentacles knot. We are all exposed to songs and stories full of ageist caricatures; we all interact with a health-care system that medicalizes aging; we're all awash in a sea of ageist pop culture. Most of us will experience cumulative ageism in multiple realms. And we know from research that the types of stress with the most severe impact on health are those that are reoccurring and unpredictable, which is the chronic and erratic way that ageism is often experienced.[94]

Ageism is compounded by sexism, racism, homophobia, and other prejudices people are exposed to throughout their lives. In the US, people of color and women are more likely to work in low-wage workplace conditions that negatively impact their health. They therefore enter old age with more health problems, less savings, and fewer options for health care. People usually don't age out of inequities; these get compounded in old age.[95]

In 2021, the US saw an alarming spike in anti-Asian violent hate crimes across the country. Asian people were physically attacked in the street, in front of their homes, or while walking to church. Many of these victims were older women, but when people spoke out against the spate of violence, there were few mentions of age or ageism. An exception was Karlin Chan, a community advocate in New York's Chinatown: "It has our seniors and the women more concerned. It seems like they're picking on seniors. These people are opportunists. They're not going to pick on a fit young man."[96]

That's an example of how ageism can be intensified by other isms. "Intersectionality" means that ageism combines with other forms of discrimination to exacerbate disadvantages and amplify their impact. Within the US, the percentage of older persons unable to afford sufficient food is highest among people of color, with 64 percent of Black elders and 74 percent of Latinx elders living just above the poverty line.[97]

Take as another example the health struggles and crushing poverty faced by many older Native Americans. Members of this group died at a high rate from COVID-19 in large part because they had limited access to health care.[98]

The compounding effects of stigmas on older Native Americans were described by Deborah Miranda, a sixty-year-old poet and member of the Chumash tribe: "It's a kind of never-ending struggle. You never get a break. There's a lot of trauma, a lot of stress." She told me that due to structural discrimination limiting access to a livable wage and adequate health care, many members of her tribe don't make it to old age, and those who do often suffer from the combined afflictions of racism and ageism and, for women, sexism.

Her grandmother died in middle age. Her grandfather Tom, on the other hand, who lived until he was seventy-five, suffered from racism most of his life. In old age this was compounded by the negative age stereotypes of the white culture that he assimilated. Feeling that he had little to contribute as an older person or as a Native American, he never shared his knowledge of tribal customs, dances, or language with his grandchildren. Deborah later found out that he would secretly "make his own regalia and go up into the mountains to dance." She didn't know that he knew how to speak the tribe's language. "It was only on his deathbed" that she first heard him speaking Chumash.

Where Do We Go from Here?

The pervasiveness and depth of ageism means that overcoming it must occur on two levels: first, confronting the negative age beliefs on an individual level when we encounter them; and second, confronting the societal institutions that operate on the basis of those beliefs. Guidelines for these confrontations, which together can help establish an age-inclusive and just society, are presented in the next two chapters and elaborated on in Appendices 1 through 3.

9

Individual Age Liberation:
How to Free Your Mind

Although age beliefs are assimilated and reinforced over our lifetimes, they are also malleable. There is nothing fixed or inevitable about age beliefs: I have changed them in the lab, they can shift across history, and they can vary dramatically from one culture to another.

In this chapter I will show you how to shift from an age-declining mindset to an age-thriving one. To do this, I'll present the ABC method, which I developed for this book, based on scientific findings and observations. The method consists of three stages: increasing **Awareness**, placing **Blame** where blame is due, and **Challenging** negative age beliefs. This method will show that negative age stereotypes are not a fortress with a moat around them that cannot be breached. The strategies can help chip away at negative beliefs and reinforce positive ones. These are the goals of this chapter and the exercises that are presented in Appendix 1.

The ABCs of Age Liberation

A: Increase Awareness

Awareness Starts Within

Success in changing our negative age beliefs hinges on our ability to identify them. We can't improve our age beliefs without first taking stock of them. Monitor your age beliefs by checking yourself for portrayals of older people that feel like negative stereotypes; classify these portrayals as such. If you find yourself behind the wheel, muttering under your breath about the elderly driver in front of you, remind yourself that older drivers have fewer accidents than younger drivers and are less likely to text while driving.[1] You can also think about the many excellent older drivers, such as NASCAR's Morgan Shepherd, who raced at the age of seventy-eight.

It helps to be aware of how we speak to older persons. In the US and Europe, when we speak to older people, particularly those who are receiving care, many of us employ "elderspeak," which involves using simplified language, a singsong cadence, and a louder-than-normal voice.[2] We also sometimes call old people names normally reserved for children or puppies, such as "cute," "cuddly," "dearie," or "sweetie." This kind of language can easily lower the recipient's feelings of self-worth.[3] Recently, when talking to a centenarian, I found myself speaking loudly and using predominantly monosyllabic words. I quickly realized she had no trouble hearing or understanding me; she was even looking at me with a sly little smile, like she knew what I was doing. So I deliberately adjusted my style of speech to how I would speak to a close friend of my age. Before I knew it, I was using my normal, everyday language with her again.

Awareness through a Portfolio of Positive Images of Aging

The more we become aware of and absorb positive models of aging, the more our conscious or unconscious negative age beliefs, that we have assimilated from the ageism around us, break down.[4] Think of someone you see as a positive model of aging, such as a parent, your neighbor, your college history professor, your reference librarian, Kusi the Zimbabwean grandmother who offers talk therapy on the Friendship Bench, the sixty-something barista who cracks hilarious comments about current events. How does this person's behavior disprove a negative stereotype or strengthen a positive one?

Positive role models don't just make us feel good; they actually help change our behavior. Take the "Scully effect," named after the fictional FBI scientist Dana Scully, played by Gillian Anderson on *The X-Files*. Girls who grew up regularly watching her were more likely to study science and enter a scientific field.[5]

There are other ways it helps to have positive models of aging. I found that older people who briefly journaled about an imagined day in the life of a hypothetical healthy, active older person once a week for four weeks significantly reduced their negative age beliefs.[6] Several studies by others have found consistent results. People who grow up with older role models in the home before the age of *two* were generally healthier in later life than their peers who went through early childhood without such role models.[7] And college students who were primed in experiments with older role models, such as Mother Teresa or Albert Einstein, scored much lower on implicit ageism than the college students who were not.[8]

In addition to Grandma Horty, I grew up with other grandparents I admired and adored, and parents who continue to inspire me in their old age. As I write this, my seventy-eight-year-old mom, Elinor, a passionate immunologist who ran an innovative medical research lab, heads a chapter of Grandmothers in Action, a group that organizes Get Out the Vote campaigns. My eighty-five-year-old dad, Charles, a sociologist whose research with Vietnam veterans laid the groundwork for identifying PTSD, now works tirelessly as an adviser to younger researchers (including me).

It's important to build up a diverse and nuanced portfolio of positive images of aging. In that way you can associate different kinds of admirable qualities with aging.

It may be counterproductive to model your life after only a single image that is exceptional or *too* positive; for instance, John Glenn, the astronaut-turned-senator who went back into space at age seventy-seven or Supreme Court Justice Ruth Bader Ginsburg, who wrote brilliant court opinions into her late eighties, because these potential role models allow us to label them as exceptions.[9] After all, how many of us switch back and forth between two high-flying careers, or serve on the highest court in the land? That said, though, noticing specific qualities of older role models that we admire (such as Justice Ginsburg's work ethic, or her commitment to gender equality) is more helpful, since strengthening these qualities is a more attainable goal for most of us.

Awareness of Age Diversity and the Impossibility of Age Blindness

Aging is an especially heterogeneous process: in fact, we become more different from one another the older we become.[10] This is due to both societal and individual factors. Thinking of everyone over the age of sixty as the same makes about as much sense as lumping everyone between the ages of twenty and fifty in the same category.[11] Unfortunately, many news stories and health studies in the US and globally either exclude older people or place them in a homogeneous demographic. This makes it impossible to take a closer look or to create policies and programs that could better direct resources to this age group. It also makes it easy for us to avoid considering the remarkable diversity of the aging process.

Just as color blindness dismisses the importance of race, age blindness dismisses the importance of age. If you look out for this, you will start to notice how common it is. A well-intentioned man who sells fish at my local supermarket calls the older customers "young lady" and "young man." In the US, it is common to say to adults we haven't seen for a few years that they look like they "haven't aged a day." Although this is intended as a compliment, ignoring or downplaying someone's age can be demeaning to that person as it suggests that age identity should be minimized.[12] The best approach, therefore, is not to pretend that aging doesn't occur. Being older is something to take into consideration and to value. Pretending not to notice it sweeps the advantages, as well as the discrimination that can accompany it, under the rug, and therefore isn't much of a solution at all.

Awareness of Invisible Age Stereotypes in Everyday Life

In addition to looking within and studying your portrayal of others, look for age stereotypes everywhere else. At first, it may feel like you are searching for something that's invisible. It's like the joke about two young fish who are swimming along when they meet an older fish who says, "Hey, guys, how's the water?" The two young fish continue to swim along, when one finally says to the other, "What the hell is water?"[13]

Once you start paying attention to the water, you'll see that everywhere you look is wet. In my Health and Aging class at Yale, students start the semester with relatively little awareness of ageism; three months later, they can't pick up a newspaper, look at social media, or talk to others without noticing the negative age stereotypes lurking everywhere in their lives.

One student was suddenly shocked by an airport security sign she had seen many times before: "If you are under 12 or over 65, you do not need to take off your shoes." She had never given much thought to why the Transportation Security Administration, a federal agency, equates these two age groups, or to the impact this infantilizing might have on older people.

Something called the Baader-Meinhof effect[14] takes place when you start to pay closer attention to a single phenomenon. Say you're thinking of buying a new car; for instance, a Subaru station wagon. Suddenly, they're everywhere: you notice them on the highway, in the airport parking lot, on your street. It turns out your friend's sister drives one; you learn that it was your father's first car. It seems like a conspiracy, but it's not. It's simply that the Subaru station wagon is on your mind, so you notice it more often. It's the same thing with ageism: once you start thinking about it, you will see it everywhere, in nearly everything and everyone.

Some forms of ageism are easy to observe: for example, the aisle that my local party store reserves for older birthdays. If you can't find it, look for the sign that reads: "OVER THE HILL." There, you'll find black balloons with tombstones on them, and tablecloths with grim warnings printed on them: "If you were a horse, you would have been shot by now."

Other instances of ageism are more challenging to notice because they involve the absence of older persons. Examples include a hospital that denies a needed medical treatment due to the patient's age; not including age in discussions about promoting diverse representations in media, marketing, and the workplace; and denying older persons inclusion in medical trials that could lead to improved treatments. Even though these are easy to overlook, it is important to look out for whether or not there are equal opportunities for and inclusion of older persons.

Awareness of Our Future Selves

For those of us who aren't yet old, instead of viewing ourselves as fundamentally different from older people, it's helpful to think of ourselves as **older people in training**. If all goes well, we will become old. In this light, your negative age beliefs can be recast as a prejudice against your future self.

It's difficult, when you are young, to accurately picture being older, especially if you don't have close contact with someone who is older. Resisting old age and avoiding older people is one way many young people learn to act, often without their awareness, in an ageist society. A better way of approaching aging is through active intergenerational contact, which is a win-win. Seek out older people, be it in the form of intergenerational yoga classes, an online book club, public spaces that welcome all ages, or an age-justice group like the Gray Panthers, whose motto is "Age and youth in action." Get to know an older colleague or neighbor. Plan a project with an older relative. A recent global survey of studies that looked at the impact of bringing together adults of different ages for various activities (such as volunteering in a soup kitchen) found that this improves younger people's views of older people and vice versa.[15] If direct experiences aren't readily available, seek opportunities for greater exposure to movies, books, blogs, podcasts, and other media created by older people.

B: Place Blame Where It Belongs

Blame Ageism, Not Aging

Once you've become aware of your own age beliefs and of those permeating your culture, you're ready to start reshaping your understanding of aging with the B phase of the ABC method. This involves redirecting the blame from oneself when targeted by ageism, including negative age stereotypes, to the proper target of blame: the ageism itself and its societal sources. This requires looking at the larger context in which a bad event occurs to find the true source of the problem. A good place to begin is by recognizing that it is often ageism that makes being old difficult, not the aging process itself.

I recently heard a doctor tell the story of an eighty-five-year-old man who goes to see his doctor with a dull pain in his knee, only to be told, "Look, this knee is eighty-five years old. What do you expect?" "Well, yes, Doctor," the patient replies, "but my other knee is eighty-five years old, too, and it doesn't hurt one bit."

A doctor dismissing his patient's concerns because the patient is old means the doctor is blaming old age when something else could be at play. By failing to investigate the source of the problem and relying instead on the ageist assumption that decline in later life is inevitable (an age belief that is often implanted in preschool and reinforced all the way through medical school), this doctor is shirking his medical responsibility. It could be that the patient's knee is bothering him because he recently pulled a muscle while shoveling snow in his driveway. But since it's "an old knee," the doctor thinks the problem isn't his to deal with. He's misguidedly blaming the problem on something that is supposedly inherent to aging and therefore can't be helped.

We have a natural tendency to blame people rather than their situation when something goes wrong. This is called the fundamental attribution error.[16] If someone cuts you off while you're waiting in line at the cash register, you might assume that they're rude, rather than a harried parent who is rushing to buy medicine for a sick child at home.

Frequently, when I present my research to audiences, people come up afterward to say something along the lines of, "Well, of course negative age stereotypes abound. They are describing the reality that debilitation happens hand-in-hand with growing old." The first problem with this idea is that the popular narrative of aging as a time of inevitable mental and physical decline is incorrect. Recall Patrick who continued to expand his vast recall of mushroom species in his seventies or Maurine who became a competitive swimmer in her nineties. The second issue is that this line of thinking gets cause and effect mixed up. As described earlier in the book, societally based age beliefs influence our health and the biological markers of aging. When it comes to how we age, society is often the cause, and biology the effect.

So what's the best way to reframe our causal thinking? We can shift the blame.

Blaming Upstream Causes: Saving the Drowners

You might wonder why I'm suggesting that we reshape our age beliefs to improve health rather than modify concrete health behaviors. After all, most books on aging health recommend focusing on things like eating well, reducing stress, and exercising. Although these behaviors are all helpful for health and longevity, targeting these types of behaviors is often unsuccessful in the long term, and sometimes even counterproductive.[17] Why? When you're focusing on poor food consumption, high stress levels, or lack of exercise, you're dealing with *downstream* factors rather than the *upstream* ones. Let me illustrate what I mean with a parable from the medical sociologist Irving Zola:

You're standing on the banks of a fast-flowing river when you see someone struggling in the water—perhaps even drowning. At the risk of your own life, you dive into the freezing and dangerous current and manage to bring that person back to shore. You're performing CPR when you hear a scream and see another person struggling in the water. You dive in once more and hoist them to safety and begin resuscitation again, when the current carries forth several more drowning, gasping people. An awful panic seizes you as you realize something is happening farther upstream causing all these people to fall into the water. Your impulse is to try to prevent people from falling into the water by intervening at the source, but you can't investigate or do anything about it when you're too busy saving the drowning people.[18]

That is a classic challenge of public health: the dual need to address the pressing, urgent downstream problem (drowning people) and the dangerous, structural upstream cause (whatever it is that's causing people to fall into the water). Age beliefs are an upstream predictor of health and well-being. Improving your age beliefs allows you to change your health habits more easily than focusing on these habits alone. Belief-based habits can be changed from the inside out.

In the parable, the upstream factor of ageism, which includes negative age beliefs, could be represented by a villain who is pushing people into the river. We need to restrain the villain. The ideal correction is a wave of social reform that would sweep away ageism. Until then, however, we can protect ourselves from the choppy waters with the ABC exercises outlined in this chapter and described in more detail in Appendix 1.

C: Challenge Negative Age Beliefs

The C of the ABC method stands for challenging age beliefs. We have found in our research that older people who actively cope with ageism by confronting rather than ignoring it are less likely to develop depression and anxiety.[19] Specifically, this applies to negative age beliefs: the more we challenge them, the less firm their grasp is on us.

Call It Out

Challenging ageism means calling it out when you see it. That applies to private interactions and public forums.

For example, sixty-four-year-old antiageism activist Ashton Applewhite started an online column called "Yo, Is This Ageist?" in which readers ask her if she considers various words and deeds to be ageist. Applewhite calls out ageism in gentle, thoughtful ways and encourages her readers to call it out as well. Recently, one reader asked: "What are your thoughts on the phrase 'for the young at heart'?" Applewhite's reply: "What does 'young at heart' mean?

Playful? Romantically inclined? Open to adventure? We can feel any of those things—or their opposites—at any point in our lives. Youth-centric language like 'young at heart' is ageist because it suggests otherwise."[20]

Increasingly, celebrities are speaking out against ageism. For example, megastar Madonna recently complained, "People have always been trying to silence me for one reason or another . . . and now it's that I'm not young enough. Now I'm fighting ageism, now I'm being punished for turning 60."[21] Similarly, Robert De Niro, at the age of seventy-two, complained about the movie business: "Here, youth is a very important part of the culture. Age is not as revered, if you will, as it is in other places. . . . To be in a movie, they would prefer you to be young and pretty or young and handsome."[22]

Not-so-famous people, too, are taking matters into their own hands. My eighty-three-year-old father-in-law, for instance, is a popular music professor and accomplished Juilliard-trained pianist who filed an age-discrimination case against his employer, a Pennsylvania university, that tried to force him to retire by making his day-to-day life increasingly difficult.

Challenging can start early. When I was raising my daughters, I tried to expose them to positive age beliefs by reading them books that featured compelling and interesting older characters. And I chose TV shows that featured a diversity of ages. I also tried to protect them from absorbing negative age beliefs by encouraging them to point out examples of ageism when they slipped in. One Roald Dahl novel, *George's Marvelous Medicine*, surprised them by opening with a rather unfortunate description: "That grizzly old grunion of a grandma had pale brown teeth and a small puckered-up mouth like a dog's bottom." One of my daughters frowned when we read that. My other daughter said, "Well, that seems mean to the grandma."

Another form of challenging also arose when my daughters were in elementary school. Predating the hundred-day celebrations that mock older persons in American elementary schools, they participated in a talent show that featured MCs, two popular fourth graders who donned white wigs, ratty slippers, and canes to shuffle around the stage with pronounced stoops and a grumpy attitude. The MCs would announce an upcoming act and wander off mid-sentence, as if they had forgotten what they meant to say. The audience, including many teachers, laughed and cheered.

To challenge the ageist caricatures of the MCs, my daughters and I created and carried out an antiageism workshop where the kids could talk about the age stereotypes they had observed on-stage and older people who were different from these portrayals. My daughters also led their classmates to work in groups to create collages from pictures that either came from a pile of magazines or from their own experiences. Half the groups created collages to show ageism and half the groups created collages to show positive portrayals of older persons.

One such older person, who doesn't hesitate to challenge age-ism, is ninety-nine-year-old Irene Trenholme, one of my parents' neighbors in the Green Mountains of Vermont. She runs a used bookstore, called Secondhand Prose, that donates its profits to the local library. She schedules all the shifts of the bookstore workers and sorts the donated books. On a recent afternoon, Irene invited me for coffee in her sprawling Victorian home on top of a hill. On the coffee table between us sat a thousand-piece jigsaw puzzle of a Gustav Klimt painting that she had just completed—she slides and snaps together the pieces of a new puzzle every week to stretch her mind and fingers.

Irene told me that when she encounters ageist remarks or be-havior, she politely but firmly points out that it isn't appropriate. Recently, her doctor was speaking loudly to her, even though she doesn't have any trouble hearing, and addressing the bulk of his remarks to her son. Irene found it necessary to point out that "my hearing is fine, thank you. And my son is not your patient, I am." The doctor, somewhat embarrassed but nevertheless gracious, told her he hadn't realized how he'd been acting.

Irene also steps in when her friends are the butt of ageist jokes or behavior. "People are not always kind when we get old," she says. So she does her part to try to change that. Her bravery, she told me, comes from her own grandmother, who raised her on a dairy farm on the outskirts of the same small town where she now lives. "Whatever you do," her grandmother used to tell her, "I want you to stand on your own two feet."

In Appendix 1, you will find a set of exercises that will allow you to carry out the ABC method of bolstering positive age beliefs. To exemplify what the ABC method looks like when applied to a real-world setting, let me introduce you to Susan.

From Papua New Guinea to Change Agent: ABC Method Personified

At the University of Chicago, Susan Gianinno, then a young doctoral student, was facing a quandary. She had been getting ready to leave for Papua New Guinea to conduct months of extensive field research on the local social dynamics when her baby daughter was diagnosed with chronic tonsillitis—nothing major, but serious if untreated. The baby would need close medical supervision and antibiotics, neither of which were easily accessible at the time in the South Pacific island.

She decided that instead of traveling halfway around the world to conduct her dissertation, she would find a way to complete it in the United States. While figuring out how to adjust her dissertation topic, she got a phone call from the head of research at a large ad agency. He had been given her number by a professor at the university and was hoping to recruit social scientists to give his firm an edge. Susan hesitated for a few days before taking the plunge. She accepted the advertising job.

"I was placed on the McDonald's account," she says, laughing. Within weeks, she went from her plan to study how social support factors influence well-being on an island in the South Pacific to trying to figure out "why the Burger King's Whopper was suddenly outselling Ronald McDonald's Big Mac." It was an adjustment. Her new colleagues dressed well and paid lots of attention to self-presentation ("In academia," she jokes, "if you tuck in your shirt, people think you're being fancy"); things moved fast: projects were completed in weeks (at the University of Chicago, you could spend a decade laboring on the same project). Susan found her new world dizzying and thrilling.

Awareness of ageism in the ad world came to Susan early and vividly. One of her first clients was the skin-care brand Olay. The first Olay ad she saw come out of her firm showed a young woman, ostensibly not yet an Olay customer, staring at her reflection in the bathroom mirror where she sees an older woman looking back at her. "I look just like my mother!" the young woman shrieks.

Older people were frequently absent from the firm's ads; when they were included, they tended to be portrayed as docile grand-mothers, if older women, and domineering grouches, if older men. And generally, they were portrayed as being out of touch with tech-nology. Fewer than 5 percent of ads featuring older people show them handling technology, even though 70 percent of people be-tween the ages of fifty-five and seventy-three own a smartphone.[23]

In addition, Susan has noticed the portrayals of older people by her industry are only getting worse. Even though people in their seventies are the fastest-growing segment of the American work-force, they're rarely featured in ads that portray people at work. Instead, they often appear as recipients of medical care.

Over the last decade, Susan says, "marketing has been blasting the airwaves with ads aimed at older people." This is somewhat ironic, she points out, "since recognizing them as a market segment with spending power could be seen as a sign of progress. Unfortunately, the deep theme in most of these ads is that aging is a time of debilitation, decline, and problems. The issue isn't that loneliness or diabetes don't exist in later life, but that there's no counterpoint to it. Older people just get bombarded with these negative messages. Not the flip side, not the broader view, not the multidimensional view of aging."

Susan **blames** the ad industry for this view, since these age stereotypes don't square with what Susan viscerally knows to be true about the diversity and strengths of aging. She was raised in a large intergenerational family, with eight siblings, an active, working mom, very involved uncles, aunts, and grandparents, and a dad who worked as a doctor until he was well into his nineties. And at the University of Chicago, Susan studied with the founder of social gerontology, Bernice Neugarten, who broke myth after myth about aging by showing, for instance, that the "empty nest" phenomenon of grown children leaving home, while portrayed in popular culture as a sad or even traumatic experience, is often instead an opportunity for growth for both the child and the parent.

Ads reflect the beliefs of the people the agency hires to create them: the average age of employees at advertising firms is thirty-eight.[24] In Susan's own firm, executives frequently staffed client-facing teams based on how young employees looked, and older employees were often excluded from participating in the creation of ads. Without a seat at the table, their voices were excluded.

Susan decided that if she was going to stay in advertising, she needed to shake things up and **challenge** commonly held assumptions. As she moved up the many rungs of her hypercompetitive industry, she worked on reframing its practices, with the goal of embracing the diversity of aging, both in ads and in the conference room.

Today, Susan is the chair of Publicis North America, one of the largest advertising companies in the world. Now in her seventies, she is one of its oldest and most powerful CEOs. While she was quietly transforming the ad industry from within, she also turned to the nonprofit world, where there were less constraints and fewer profit motives standing in the way of new and better portrayals of aging.

Today, Susan helps run the nonprofit Ad Council, which produces advertisements for public interest causes, such as the famous Got Milk campaign. She is particularly proud of the Love Has No Labels campaign she spearheaded, which tackles all kinds of prejudice by showing many different types of people in loving relationships.[25] These people are old, young, gay, straight, Black, Brown, white. In one of the ads, a large x-ray screen is set up on a boardwalk. Different couples hug behind the screen, which conceals their identities, revealing only their interchangeable skeletons. One by one these couples reveal themselves. The last is an older man and woman. As they emerge from behind the screen, the woman proclaims, "Love has no age limits." The ad has been watched by over sixty million viewers.

On a personal level, Susan has been challenging negative age stereotypes by engaging in what she calls "non-stereotypical interactions" with her own large and close-knit multigenerational family. Recently, for instance, they all learned to paddleboard together. Susan's young grandson gave her advice on how to integrate a new technology into one of her firm's campaigns, and Susan instructed her young granddaughter on how to call a board meeting to order and then let her do so.

Meanwhile, Susan told me that she is convinced that change is around the corner: "Eventually, the sheer numbers of engaged, healthy, happy, active older people who are fighting for a more just world will do the work for us." She expects the catalyst to come from what she calls "norm entrepreneurs: They're not traditional activists, and they seem like rabble-rousers until there's more and more of them, and suddenly the numbers snowball and they acquire velocity, and what seemed like noise is now breaking out into mainstream culture."

Societal Age Liberation:
A New Social Movement

The Rally to End Ageism

On an unseasonably dreary recent April afternoon, I took the train to New York City with two of my grad students, Samantha and Iggy, to stand in a packed and soggy crowd just outside Central Park. We were there to join the first-ever Rally Against Ageism, which included people of all ages, races, and backgrounds—from toddlers in strollers to activists in their nineties. The gathering confirmed what everyone there already knew: that society is a web of interdependencies—what affects one group affects all the others. Old age was on our minds and the injustice of ageism was weighing down our hearts, but the joyous sight of this cross section of society gathered in solidarity and protest was something I will not soon forget.

The afternoon was full of rousing speeches and chants. Local councilwoman Margaret Chin delivered a stirring defense of elder rights: "We have to change that dangerous narrative that projects seniors as a burden on society. We've got to turn that around!" Several New Yorkers who experienced age discrimination at the hands of doctors, landlords, or employers told their stories. Homemade signs competed for the real estate above our heads (mine featured a picture of Albert Einstein; underneath, I'd written: "Would you hire this man?"). There were exhilarating dance performances from the Silver Sirens, a cheerleading squad of senior citizens who champion age-justice issues, and a rousing, moving aria from Black gospel singer Diana Solomon-Glover, who adapted the civil rights ballad, "Ain't Gonna Let Nobody Turn Me 'Round" to the theme of the day: "Ain't gonna let ageism turn me 'round / Ain't gonna let nobody turn me 'round / I'm gonna keep on walking, keep on talking / Marching into freedomland." When she returned to the chorus, the entire crowd was singing along with her.

The Gray Panthers: Then and Now

In 1970, a sixty-five-year-old woman named Maggie Kuhn was fired from her job at a Philadelphia church, where she had been running social-outreach programs on issues such as the fight to expand low-income housing.

After her forced retirement, Maggie got together with a few friends who had been similarly forced out of jobs for the alleged offense of being sixty-five. At first, they met to gripe, but soon the griping turned into determination. In the preceding decades, Maggie had worked in the civil rights movement and the anti–Vietnam War movement, where she learned firsthand the power of grassroots groups to effect social change. Maggie believed that although change takes time, what often seems impossible becomes, with enough passion and effort, inevitable.

Within a year, Maggie and her friends had attracted thousands to their cause, which, as she explained, was aimed at replacing the prevailing view of old age "as a disastrous disease which nobody wants to admit to having" with the then-radical notion that age was in fact a victory to be celebrated.[1]

By the mid-1970s, Maggie's little group was making enough

noise to garner national attention. One evening, a quick-witted TV newsman referred to them as the "Gray Panthers" (based on the Black Panther movement) and the name stuck.

To fight ageism, the Gray Panthers favored shaking things up through lawsuits and noisy street protests that used humor to attract attention. Their first year in operation, they dressed in Santa suits and picketed a department store the day before Christmas to protest its mandatory retirement policies, holding satirical signs claiming that Santa was too old to work there. Frustrated with the American Medical Association's lack of concern for older Americans' health issues, they dressed as doctors and nurses and made a "house call" to its convention before offering a diagnosis: the AMA "lacked a heart."[2] They protested in front of the White House to demand inclusion in the Presidential Conference on Aging. Maggie Kuhn even appeared on the widely watched *The Tonight Show* to trade zingers with its host, Johnny Carson. Overnight, she became something of a folk hero.[3]

By the time she died, in 1995, Maggie and the Gray Panthers had helped persuade Congress to reject proposed cuts of Medicare and to pass laws abolishing the mandatory retirement age in most industries.[4] In the process, they demonstrated that old age was a time for self-determination and liberation. Maggie wanted her fellow Americans to understand that "we are not mellow, sweet old people; we have got to effect change, and we have nothing to lose."[5]

A legacy of the Gray Panthers from when it was headed by Maggie is what she called its "cubs," her affectionate nickname for its younger members. Many have moved into powerful positions in academia and politics, including the founding scholar of the political economy of aging, Carroll Estes, and US senator Ron Wyden, who, at age twenty-eight, co-led Oregon's Gray Panther chapter alongside an eighty-two-year-old retired social worker and is currently one of the leading pro-age figures in Congress.[6]

When Jack Kupferman, now aged sixty-six, took over the Gray Panthers in New York, about ten years ago, the organization Maggie Kuhn had built was in a state of suspended animation. "It's what happens when the charismatic leader dies and doesn't leave an infrastructure in place. You can't maintain the same kind of attention," Jack told me, explaining the challenges of running a grassroots organization from his apartment in Greenwich Village, without a budget and with the help of only a few dozen dedicated volunteers.

Jack assumed his role with the Gray Panthers after many years as a lawyer with the New York City Department for the Aging. In one way or another, he has spent his entire adult life fighting for the dignity of older people. When he was growing up, his parents ran what would today be called an assisted-living facility out of an old converted farmhouse in upstate New York. Its residents, like the retired opera singer who taught him to sing scales, were part of his family.

When Jack went to college, he knew he admired older people, but he wasn't sure how to include them in his life. One day, he happened to see white-haired Maggie Kuhn speaking passionately on TV. "She was just the bomb. She was like, 'Excuse me, why is there mandatory retirement?' Ageism is a problem not just for older people, but a social justice problem. We need to make change not just as a kumbaya-nice-to-hang-out-together issue, but to make a better world." So Jack went to law school.

Since then he's established a literacy program for older people in Nepal and a microfinance fund for older women in Pakistan. More locally, he was the catalyst for an investigation by the New York State comptroller into nursing homes with bad records and ran a task force after Hurricane Sandy devastated New York City to ensure that the needs of older New Yorkers were included in emergency relief. That effort started when he visited an evacuation shelter in Brooklyn and saw hundreds of disabled older people who had been piled into a gymnasium. They were the residents of an assisted-living facility that had been destroyed by the hurricane. Jack learned that the owners of the business hadn't even called to find out how their residents were doing. The shelter didn't provide soap or enough food. Jack was furious. His fury returned with the recent and similar neglect of older persons at the start of the COVID-19 pandemic.

Launching an Age Liberation Movement

To imagine the possibilities for an age liberation movement, I looked at other social movements that successfully shaped American cultural norms. The LGBTQ movement, for example, changed most Americans' attitudes about same-sex relationships in a short period of time. As recently as 2004, two-thirds of Americans opposed same-sex marriage. Today, two-thirds of Americans support it; this support has increased among all demographic groups, across all generations and religions and political affiliations.[7]

As I was thinking about social movement strategies, I called up my dad, a sociologist, to glean some insights. Early in his career, he moved to the South to help support the civil rights movement and teach at Tuskegee Institute, a historically Black college in Alabama, where my brother and I spent the first years of our lives. During that time, he contributed to the movement through his investigative reporting and fundraising.

I also drew on my research findings of the best strategies for encouraging beneficial change. From these diverse sources I identified three stages that could lead to the formation of an age liberation movement and the successful achievement of a society that protects the rights of older persons: **collective identification, mobilization,** and **protest**. (Also see specific strategies to combat structural ageism in Appendix 3.)

Age Liberation Movement Stage One: Collective Identification

Collective identification involves creating a sense of belonging to a group by making its members aware that they are targeted by age- ism and by helping them to see it as caused by societal forces that can be changed. The goal for this phase is to instill what sociologist Aldon Morris has called "cognitive liberation,"[8] which occurs when people collectively decide to resist stigmatization.

A central aspect of collective identification involves articulating grievances and combining these with research-backed evidence of the widespread societal damage caused by the problem. This was a crucial component of the Black Lives Matter and Me Too move- ments, when powerful and often horrifying individual stories of racial and sexual assaults were amplified and made even more im- pactful by convincing data about the entrenched prevalence of the underlying problems: racism and sexism.

Collective identification played a key role in raising group con- sciousness that led to the women's liberation movement. This started with Betty Friedan's 1963 bestseller, *The Feminine Mystique*, which grew out of a survey of her former Smith College classmates for their fifteenth anniversary reunion. When she realized there was widespread unhappiness among women of her generation, Friedan began investigating the structural and cultural foundations of what she termed "the problem that has no name." Through hundreds of consciousness-raising groups, she inspired women to share their sto- ries of individual struggles. They found commonality and solidarity in the widespread nature of their experiences.

Recently, organizational efforts at consciousness raising around ageism have been increasing. A leader in this effort is UK-based

HelpAge International, which is devoted to helping older people challenge discrimination and overcome poverty. The director of its countering ageism programs, Jemma Stovell, told me that a problem she encounters is that many older persons she meets haven't heard of the concept of ageism. That is because languages often don't have a name for it and many people aren't used to identifying the mistreatment of older individuals as due to an age-based stigmatization. Jemma often witnesses a lightbulb moment when people are given the language to finally describe the ongoing mistreatment of elderly individuals. An older man from Kyrgyzstan told her about a group of youths mocking and then attacking an older woman selling wares at a market. Before he learned the term "ageism," he had thought of the attack as simply "mean-spirited." With his new awareness, he suddenly realized that the attackers were likely motivated by ageism.

In its toolkit for raising consciousness in countries across the world, HelpAge International uses proverbs that refer to older people to show how ageism is embedded in the local cultures where its workshops are conducted. Among these proverbs are: "Useless as coconut shells" (Thailand); "Gray hair into the beard, devil into the rib" (Russia); and "Old brooms are thrown in the fire" (Germany).

Consciousness-raising has been facilitated by the internet. A group of activists against ageism in Korea posted a photo of a young person holding an umbrella, with the many protected rights of the young pasted to it. Standing to the side was an older person holding the bare wire frame of an umbrella, with bits of paper listing the many unprotected rights of the old littering the ground. This image went viral; people in dozens of other countries started sharing local versions of the photo.

Age Liberation Movement Stage Two: Mobilization

The next phase of movement building, mobilization, entails bringing members of the group together around a set of common goals that include reducing stigmatization and unfair treatment. Rosa Parks is known today as the woman who started the Montgomery bus boycotts that turbocharged the civil rights movement. Less widely known is that four months prior to refusing to give up her seat on a bus to a white man, she completed a workshop at the Highlander Folk School, a social-justice leadership training school in the Appalachian Mountains, where she learned about nonviolent civil disobedience as a tactic.[9] The ethos of the Highlander Folk School, as expressed by its founder, Myles Horton, is: "Not as individuals, but the group as a whole has much of the knowledge that they need to know to solve their problems."[10] Although his focus was on civil rights and labor issues, this insight equally applies to fighting ageism.

The internet has also benefited mobilization—in ways that were previously unimaginable. For example, Pass It On Network, an organization of older persons, now operates in forty countries as an online platform for disseminating information about aging-related issues, including ageism-based social problems.

The arts are another path available for mobilization. A group in Canada has organized intergenerational dance flash mobs in public spaces, such as shopping malls, with participants ranging in age from fourteen to ninety-two slowly joining together to perform the same dance movements in unison. Their goal is to unite dancers and create a dance performance that "shatters commonly held stereotypes about aging," as one of the dancers described it.[11] There is also the Theater of the Oppressed, pioneered by Brazilian activist Augusto Boal, in which spectators become actors, first witnessing an acted-out example of prejudice before then participating in addressing it onstage. In one performance, called *The Runaround*, the audience is encouraged to challenge age injustice in the health-care system after watching an older character be denied urgent dental treatment by her insurance provider on the grounds of being too old.[12]

Satire can be another effective way to mobilize people. In my personal favorite, *Gulliver's Travels*, Jonathan Swift satirized numerous aspects of British society, including its ageism: he invented a land populated by mythical beings, Struldbrugs, who never die, but are stripped of rights, property, and dignity the moment they turn eighty. Even though Swift wrote *Gulliver's Travels* three hundred years ago, when I read the book each year with my students, I'm always struck by how resonant his message is today.

In a more recent example of ageism satire, this time in the context of the movie industry, the comedian Amy Schumer, who was in her thirties at the time, stumbles into an outdoor party for Julia Louis-Dreyfus, who, in her fifties, is celebrating her last day of being sexually desirable before being put out to sea in a boat.[13] As she explains to Amy, "In every actress's life, the media decides when you've finally reached the point where you are not believably f***able anymore." Aghast, Amy asks, "How do you know? Who tells you?" Tina Fey chimes in: "Nobody really overtly tells you, but there are signs." Julia continues. You go to a movie set, you go to wardrobe and the only thing they have for you to wear are long sweaters, like, cover you up head to toe kind of thing. In one five-minute skit, these actresses managed to tackle ageism in Hollywood, in the fashion industry, and in our culture's tendency to desexualize older people. It has been viewed online nearly seven million times.

Age Liberation Movement Stage Three: Protest

The final stage of any effective social movement, protest, involves its participants directing their energy at the structural sources of their marginalization to spark social change.

A successful age liberation movement could be backed by the formidable power of older voters. In addition to representing an increasingly significant swath of the American population, older people historically have the highest proportion of voter turnout.[14]

Drawing on such political clout, an age liberation movement could, for example, demand a government-backed public-information campaign against ageism. It could follow the model of the one that was directed against smoking, which had considerable success—not only in the US, where it originated, but also in countries ranging from the Netherlands to New Zealand.[15] Just as the anti-cigarette campaign had as its centerpiece "Smoking is dangerous to your health," a new public-information campaign could warn, "Ageism is dangerous to your health." It could shed light on the toxic impacts that ageism has on the range of cognitive and physical outcomes, as documented by my team's research. A broad range of outlets could be used, including social media, television, and print media.

An age liberation movement could also wield its power in the private sector. In the US, the over-fifty population is responsible for the majority of consumer spending, a trend that is spreading worldwide.[16] In the UK, for example, this older age group was responsible for 54 percent of consumer spending in 2018, a number that is projected to rise to 63 percent in 2040.[17]

Although there is a wide range of appropriate targets, the advertising industry is the one that does the most to promote ageism, given that it underwrites so much of traditional and social media. The specific goal, then, would be to end the demeaning presentation of the old and to demand the inclusion of positive and diverse images of older people.

Other targets of protest could be television and social media. TV has a twofold problem: the ads themselves are often ageist, and the characters in shows tend to be based on negative age stereotypes.[18]

Social media sites, as our research has shown, provide another forum for disparaging older persons.[19]

The first step in addressing this problem would be discussions with advertisers about the pernicious health consequences of practices that foster negative age beliefs. If the discussions do not lead to change, boycotts of the relevant media platforms and the companies that advertise on them could follow.

In July 2020, over a thousand companies joined the Stop Hate for Profit campaign, threatening to boycott Facebook unless the social media giant stopped disseminating posts on a range of hateful or dangerous topics, from racism to voter misinformation. A diversity of celebrities, such as Katy Perry and Sacha Baron Cohen, supported the boycott. The campaign achieved some success: Facebook announced the creation of a team to study and prevent algorithmic racial bias.[20] None of the demands, however, covered ageism—which is another indication that older persons need to have their own social movement.

On the international level, an age liberation movement could resuscitate a worthy United Nations initiative: the "legal instrument to promote and protect the rights and dignity of older persons," which includes combating ageism. Unfortunately, a majority of the UN member states have rejected it.[21]

A few years ago, when I attended a working group of 194 United Nations member countries to discuss this convention on elder rights and how to give it enforcement capabilities, I heard impassioned testimony by older people on the need for urgent relief. I learned that 60 percent of older people in low- and middle-income countries reported they could not obtain needed health care because of their age, and that in a survey of 133 countries, only 41 had national laws to prevent violence, abuse, and the neglect of older persons. I also listened to the speakers describe how, in many countries, including the US, people are more likely to have to choose between buying medications or food as they age: among older Americans, poverty rates increase from 7.9 percent for those aged 65 to 69, 8.6 percent among those aged 70 to 74, 9.5 percent among those aged 75 to 79, and 11.6 percent for those over the age of 80.[22]

Nevertheless, the American delegation refused to ratify the UN convention. At lunch, I managed to seat myself next to the diplomat representing the US and asked if I had correctly heard her speech. It turned out I had. Between bites of her baked potato, she repeated there was no need, in her official eyes, for the convention to protect older persons, since they were already covered under an earlier disability convention. I was flabbergasted. To conflate the rights of older people with those of people with disabilities made no sense. Nor did rejecting troubling international and national data. Although protecting older persons with disabilities is essential, it is no less important to protect the rights of older persons as a whole.

Intergenerational Coalition

Any effective age liberation movement would have to be led by those most affected by ageism: older people themselves. The disability rights movement is famous for its slogan, "Nothing about us without us," which not only captures the desire to overturn the marginalization of people with disabilities, but also emphasizes the need for self-determination.[23]

At the same time, an age liberation movement would ideally be intergenerational. Younger people, who often aren't aware of discrimination toward the old, might not view it as a burning issue. But the mobilization process should convince younger people that it's not just their parents or grandparents who are victimized by ageism; it is also their future selves. In addition, younger people have become increasingly aware of the importance of social justice.

Many have joined movements that bring together people of different ages such as Sunrise, which focuses on climate change. This background makes the young exemplary potential allies.

A virtue of the intergenerational approach is that younger people could contribute a useful perspective for locating areas of ageism that need to be addressed. This was the case with Rachella Ferst, a college intern Jack Kupferman recruited to help the Gray Panthers one summer. After growing up in Singapore, when Rachella moved to the US for high school she was dismayed to find that many people talked about older persons in a negative way. She wondered if this could be due to older persons being excluded from the educational curriculum in her new home.

Rachella credits her commitment to elder justice to growing up with her grandmother at home (three generations often live together in Singapore) and to her country's school curriculum, which integrates older people in a meaningful way. From ages thirteen to sixteen, her class would regularly visit older people living nearby, as this was an opportunity to learn about history and to practice different languages with the older population. Most young Singaporeans speak English as a first language, but learn Malay, Chinese, or Tamil in school; these happen to be the first languages of older Singaporeans. In addition, Rachella's school encouraged students to talk to older people about their perspectives on historical events. After her summer with the Gray Panthers, she wants to develop educational content and policy that integrates the needs and experiences of older people.

An intergenerational age liberation movement would benefit the young as well as the old. Quinn, a twenty-one-year-old Colgate College football player, had never heard of ageism before his internship with the Gray Panthers. But he liked the idea of attending United Nations sessions, which was a perk noted in the internship listing, so he took a chance. What he got out of the internship was more than just talking with diplomats. It was a totally new awareness.

By the end of the summer, Quinn was noticing ageism everywhere. It was why, he realized, his grandfather had been fired from his job as an engineer at the battery maker Energizer after being told the company needed fresh ideas. Quinn explains that the Gray Panthers helped him see that this firing was ageist, because people of all ages can have fresh ideas. He also became aware of his own ageism when his grandparents or parents asked him for help with technological problems; he would fix the problem for them rather than teach them how to do it themselves, as he would with classmates, because he assumed they were technologically inept. It was a small but profound thing to realize they weren't. Now, he wants to follow in Jack's footsteps to become a full-time age liberation activist.

Jack's goal for the Gray Panthers internship program, and for mobilization more generally, is to link the generations together, "so that it becomes more about who we actually are, and not the stereotype of who we are."

Cultural Redefinition: Beautiful Aging

A successful age liberation movement is likely to generate a climate that benefits not only the institutions it targets, but also the way the movement's members regard themselves. For it would lead to a greater sense of self-worth among its members and this, in turn, could foster a cultural redefinition; that is, a new way of framing aging. This redefinition might include taking aspects of age identity that have been given pejorative connotations by society and turning them into proud, even defiant, attributes.

Cultural redefinition is particularly important for older persons because they don't have a lifelong membership in a marginalized group that has developed ways of psychologically protecting its members against the full impact of negative stereotypes. Those who become old often have to develop this protection on their own. Cultural redefinition provides a way of rewriting the age code by marshaling support of the group.

We saw cultural redefinition with the "Black is beautiful" chant of the civil rights movement and the gay rights movement claiming the previously stigmatized term "Queer." An equivalent sort of re-definition for an age liberation movement could focus on wrinkles and what they represent to society and often, therefore, to those who have them. As the forty-five-year-old actress Reese Witherspoon ex-plained in an interview, wrinkles are not just natural, they're *hard-won*: "I've had a whole bunch of experiences, and I can speak with a thoughtfulness about the changes I'd like to see in the world. I just feel like I earned that gray hair and my fine lines."[24]

My friend, Stacey Gordon, started a nonprofit called the Wrinkle Project in her midforties after realizing she was "no longer a young social worker, a young teacher, a young mother, a young person any-more. I was a middle-aged person. And I was starting to have a hard time with it. You know, my hair was turning gray, I was getting lit-tle wrinkles all over, and I was starting to feel ignored, like a lot of women." Around the same time, she realized from her social-work practice that societal ageism even seeped into family matters: "So much of my work is adult children who tell me, 'My parents need to do this or that,' without ever considering the older person's opinion."

So Stacey came up with the idea for Wrinkle Salons, for people to come together to share their experiences of becoming older. The Salons aren't just about wrinkles. Stacey chose that name for its symbolic power and to help reclaim wrinkles from the multibillion-dollar antiaging industry that generates considerable profit from ad campaigns designed to instill fears around these physical signs of aging. She recently told me that "fearing wrinkles stops us from aging well and from really being our full, authentic selves as we age. We get wrinkles and then we think, 'Oh, we're growing old,' and that's the beginning of what your research shows—how becoming older can trigger our internalized ageism, unless we find ways to prevent this."

I was delighted when Stacey recently invited me to help her lead the first Wrinkle Salon.

At first Stacey was planning to only invite those in middle age since they were making the transition into older age and might be particularly reflective and open to a new way of thinking. I suggested that we also invite older individuals since my research found that when aging becomes self-relevant (meaning, when we are in the thick of it) is also a good time for generating new perspectives and connections. We agreed that mingling the generations might spark a cross-fertilization of ideas across generations.

So we organized a diverse group of eleven women between the ages of forty-five and ninety-five to meet for three ninety-minute sessions.[25] During the first two sessions, held a week apart, several themes emerged. The first was that aging is often "the elephant in the room"—a big topic that is almost always on our minds but never discussed. The second was the rampant ageism most participants experienced. They spoke about being dehumanized and treated like "a dinosaur" or "a junk car that nobody wants" in their workplaces and doctors' offices.

The third idea we coalesced around was how to counter ageism and appreciate the many benefits of aging. For example, fifty-nine-year-old Alison mentioned that although she often felt dismissed by colleagues for being the oldest nurse in her hospital team, she also realizes that she enjoys being older. "I feel really good. I have raised my kids, enjoyed my career, contributed to causes I care about, and I have many goals for the next phase of my life. I feel like I have a lot of experiences I can draw on. I've seen it; I've done it; I've gone through it. Now I can help other people who are going through these experiences."

Rona, a sixty-four-year-old poet, said that she sometimes finds herself engaging in what she calls "hate talk" around aging, which includes feeling ashamed of her wrinkles. To address this, she decided she would try a new technique: imagining what a wise elder would say to her about this kind of talk. "She would probably say something like, 'That doesn't make sense. You don't need to be anxious about something just because of your age. Not being young doesn't make you less-than. Wrinkles can show experience and beauty.'"

Rona also said that the affirmation of aging that she enjoyed in the group suggested to her a shift in language that could come with aging. She asked her fellow Salon participants, "What if we shifted the meaning of the phrase 'How do I look?' from worrying about how we appear to others, to 'How do I *look?*' in the sense of welcoming aging as a time to purposefully gaze outward at the natural beauty and bigger issues of the world."

In the third and final session, everyone took a turn to discuss whether the first two sessions had changed how they thought about aging. One Salon participant, a sixty-eight-year-old therapist named Veronica, said she felt the Wrinkle Salon had shifted her thinking. She'd been "sleepwalking through ageism issues because they're just part of what you live. And then you make this space to talk and reflect and think and now you're awake and feel more aware. I just began to notice how I'm being treated. Noticing this can act as a buffer and in a certain way is protective." She explained that to undo the effects of internalized ageism, "we have to intentionally expose ourselves to alternative narratives by interacting with older people and experiencing their vitality, their curiosity, their potential. So I look for older wise women to talk to and luckily, we have some in this group!"

After some of the younger members of the Wrinkle Salon described the pressure they feel to look younger and to avoid telling people their age, the oldest person in our group, Juliet, a ninety-five-year-old retired principal, said that she now goes out of her way to tell people her age, especially when she is making an observation that she feels is helpful to a conversation. This is a way, she explained, to elevate her older identity and to say, " 'I'm smart. Don't underestimate me.' I don't want to be invisible, so this makes me very visible." She also noted, "I am much more accepting of my body now than I have ever been."

Reclaiming wrinkles is a way of countering ageism by redefining aging. As the activist and scholar Ibram Kendi writes, "To be an antiracist is to build and live in a beauty culture that accentuates instead of erases our natural beauty."[26] The same can be said for antiageism. Instead of denigrating aging with "antiaging" creams, we need to, and can, shift to a culture that emphasizes the natural beauty of all ages.

JoAni Johnson, a sixty-seven-year-old with waist-length gray hair ("a cascade of hair the colour of moonlight," a *Guardian* journalist called it),[27] is an age liberation activist who was recently hired by singer Rihanna to be the face of Fenty, her new label (in partnership with the French fashion house Louis Vuitton)—a bold move, considering that only 3 percent of the models featured in *Elle* magazine have been over the age of forty.[28] Johnson started modeling at age sixty-four, after deciding she was ready for something new: "I don't consider what I'm doing as your usual modeling career, and at five feet four inches, Black, and 67 years of age I am not your typical model in the general defined way." She credits her ninety-year-old mother, a Jamaican immigrant to the US, for showing her that beauty exists in many forms throughout the life span.[29] More than anything, Johnson values aging for what it's given her. "Being older gives me the experience of knowing that I've overcome difficult challenges before, like the death of my husband, and that gives me the confidence to face whatever comes next."

Cultural redefinition can contribute to a virtuous cycle. As individuals acquire a greater sense of their value as older persons, they are more likely to participate in an age liberation movement, and the movement is bound to further increase their sense of value as older persons.

Tipping Point to Age Liberation

An age liberation movement is an ideal, but it isn't a utopian dream. The World Health Organization, joined by 194 nations, recently launched its first campaign to fight ageism. (I am honored to serve as a scientific adviser to the campaign.[30]) In the US, the National Institutes of Health is implementing a new policy to increase the inclusion of older participants in clinical trials. Additionally, the American Psychological Association, the Gerontological Society of America, and HelpAge International have begun issuing urgent warnings about the hazards of ageism. Members of the Gray Panthers in New York City continue to look for creative ways to confront ageism.

These sporadic examples of organized resistance to ageism may provide the seeds of a movement. Desmond Tutu, the eighty-nine-year-old South African bishop, noted, "Do your little bit of good where you are; it's those little bits of good put together that overwhelm the world."

Conventional thinking holds that it takes a majority, at least 51 percent of the population, to initiate social change. But intriguing new research by Damon Centola and his team at the University of Pennsylvania shows that social tipping points can occur when 25 percent of the population decides it's time to change.[31] In other words, a determined minority can punch well above its weight. This fits in with what was known from studying sexism in the workplace: Rosabeth Moss Kanter, at Harvard Business School, showed that if a small but committed group of women pushes for change in office norms, they can successfully transform its entire culture.

Now, consider that 24 percent of the world is over the age of fifty.[32] (Even though everyone in this age group has not yet been convinced of the virtue of fighting ageism, this is the goal of an age liberation public-awareness campaign.) Perhaps all that is required for these individuals to successfully mobilize against ageism would be for an additional 1 percent of the population to join them in order to reach the 25 percent Centola identified as needed to bring about social change. That would be a tipping point. In their study of the phenomenon, the authors found that adding just a single person to the activist minority transformed their effort from a complete failure (zero converts outside of the already-committed minority) to a complete success (the entire group converted to the new outlook). By extension, a movement that appears to be a failure may actually be standing right on the cusp of success. This means that everyone, every single person who becomes aware of ageism and decides to counter it, is one person closer to a new reality.

That's what I was thinking about, on my way home from the first Rally Against Ageism. That this start of a movement has so much potential—all it needs is momentum, a swell of popular support.

Halfway through the rally, one of the organizers onstage noticed me standing in the crowd and gave me a loud shout-out. She told the crowd my research had been an inspiration for the event. All heads suddenly turned in my direction. I waved and gave a proud but embarrassed smile. I took the train back to New Haven that night in a state of amazement: that something that felt to me for so long like an uphill battle of a few might actually be the start of an impassioned movement belonging to many.

A Town Free of Ageism

Sometimes the thing you're dreaming of is right in your own back-yard. It turns out I didn't have to go all the way to Japan or Zimbabwe to find a thriving culture of positive age beliefs.

One summer day not long ago, my family and I stopped in the tiny town of Greensboro, deep in the Northeast Kingdom, a remote and hilly corner of Vermont, on the border of Canada. You may remember this town as the place where Nancy Riege makes laby-rinths. I expected we would find breathtaking views of the Green Mountains and a pristine lake full of trout and loons, and I had recently learned that one of my favorite cheeses (Harbison, a deli-cious, pungent cheese that comes wrapped in spruce bark) is made on a farm in Greensboro. Given my love of cheese and the fact that this lakeside village is on the way to my parents' house, where we were headed, we decided to visit for the afternoon.

What I didn't expect to find was a place where ageism does not exist.

We arrived midday in this town that is miles from any highway, with no traffic light and a main street that winds along Caspian Lake.[1] We stopped for coffee and sandwiches at Willey's, the town's general store, which also functions as the local deli, gas station, hardware store, coffee shop, and town square (wine, maple syrup, nails, and boots all occupy the same shelf).

While we sipped our coffees on the front porch, I struck up a conversation with a woman who had just hoisted a heavy bag of fertilizer into her truck and was now enjoying a lemonade. She was friendly and unguarded in that small-town Vermont way, and as we chatted, I found myself adopting those qualities as well. When I told her that I studied aging for a living, the woman, named Carol Fairbank, told me that I'd come to the right place.

Carol is a smiling brunette, a graphic artist in her late forties, who several years ago moved to a small farm on the outskirts of Greensboro from a large city in Massachusetts. When I asked her why she moved, she said it was because she is crazy about skiing: four months out of the year, the area is a paradise for both downhill and cross-country skiers. When she hits the slopes on clear winter mornings, "almost everyone on the hillside is white haired. And they're ripping up the slopes enjoying themselves." She also decided to move there because when she first visited, she realized Greensboro was where she wanted to grow old.

A lot of the friends she's made in Greensboro are older, and many of them are active and independent, snowshoeing in winter, stacking their own wood in the spring, gardening in the summer and fall. Carol describes how they look out for one another: those who live alone in big houses often divide them into rooms and units (with the help of a local organization that retrofits housing to keep seniors integrated in the community), which they then rent out to others—seniors and young persons alike. This allows people to age at home, if they prefer, or to move into communal housing. For those older people who are struggling financially, the town does what it can with free meals and affordable housing. And most of the art classes and cultural programming offered year-round are free or subsidized, so that all can participate. The winters can be cold, but then, Carol said wistfully as she pressed her cold lemonade to her forehead in the ninety-degree heat, "there's skiing, and ice-skating, and hot cocoa, and soup. And the sense that everyone's in it together."

Across the street from Willey's, a trio of older women was hanging a huge banner on the facade of a barn. Two of them were perched on ladders, while a third woman in a sun hat was directing them from the ground, peering up through a surveyor's scope to make sure it was straight. "Higher!" she shouted. "On the left, go higher!" The banner read: "HAPPY 100, BETH!"

Carol smiled and proceeded to tell me about one of the hottest clubs in town: the Greensboro Ladies' Walking Society, with nearly a hundred members—almost all of whom are seventy or older—who meet three mornings each week to walk together and socialize. One of their members had just turned one hundred, and her friends were hanging the banner where she would see it when she walked by.

"Men aren't invited?" my husband asked.

"They can join. But it's mostly for women. They aren't likely to get jealous, though," Carol said, "because they have their own organization." The ROMEOs—Retired Old Men Eating Out—get together once a week to have lunch at the inn next to the lake. You can join even if you aren't officially retired. Several decades ago, William Rehnquist, who had a home in Greensboro, asked to join even though he was then serving as the chief justice of the US Supreme Court. They let him in.

Carol runs programs at the Rural ARTS (Arts, Recreation, Technology, and Sustainability) collaborative, which brings residents together for classes and events. Typically, they draw enthusiastic participants of all ages, ranging from 4 to 104 years. The goal is to encourage creativity and foster connection between members of the town who might not otherwise get a chance to become friends. I later learned that Rural ARTS has become something of a model program in the region for fostering intergenerational activities. In winter, for example, there are Soup + Sustainability nights, where people get together over a bowl of soup to watch and discuss a film about environmental issues, usually led by a resident expert (often a senior, but sometimes a teen).

One of Carol's jobs at Rural ARTS is to oversee a workspace called Spark. "It's run out of a church basement," she told me, "but it's not bingo and macaroni crafts. It's very high tech: 3D printers, large-format printers, scanners, laser cutters." The space is designed to spark creativity among people of all ages. "You step in there, and it knocks a couple of stereotypes flat on their face. There are older people starting web-design businesses, printing huge banners for the local parade, just messing around with the computers and making art."

At that moment, the older woman who had been supervising the hoisting of the birthday banner approached, eating an ice cream cone. She overheard us talking about the ARTS center and wanted to share: "I use it all the time to print materials for the Historical Society. And those are *fast* printers, let me tell you: twenty, thirty pages per minute." She introduced herself as Nancy Hill, copresident of the Greensboro Historical Society. As a recently turned eighty-six-year-old, Nancy explained that she, too, had decided to enjoy old age in Greensboro. She's a fourth-generation resident who resumed living in the town after working for years in France and Thailand.

When Nancy learned that I was writing a book about aging, she also told me Greensboro is the perfect spot. She explained that Greensboro doesn't look like much of the United States, or the world, for that matter. Its population is a lot older: 40 percent of its adult residents are over the age of fifty and the median Greensboro age is fifty-two, compared to thirty in the rest of the world. And this older population is very interconnected with each other as well as the younger generation. For example, there are many intergenerational book groups and writing collectives. She added, "Our town is full of readers and writers." When she was serving on the library committee, Nancy helped to set up a shelf to spotlight pub-lished Greensboro writers. "We thought we might come up with 10 or 20 authors. Instead, we found out there have been over 150!" She started listing some of the more famous ones. There was the Pulitzer Prize–winning novelist Wallace Stegner, the Asian Ameri-can novelist Gish Jen, and the anthropologist Margaret Mead (who studied pro-age cultures in the Pacific).

"But none so famous as the famous Greensboro Ladies' Walking Society," Carol added with a wink. Nancy laughed. Years ago, after returning to Greensboro, Nancy started taking morning walks with a few friends on one side of town, while another handful of friends began walking on the other side of town. One day, the two groups ran into each other and decided to link up, and the Ladies' Walking Society has been snowballing ever since.

"It's about socializing as much as it's about walking," Nancy ex-plained. The group walks all over the Northeast Kingdom and also takes trips together—the destinations have ranged from Nantucket to the Netherlands. "Being active is important, but it's more than that. It's about . . ."—Nancy paused, trying to think of the perfect word—"being *involved*. Yes, that's a good word to describe how people are around here."

After lunch, my family and I walked around town, enjoying the fresh meadowy summer smells, the verdant fields, and the quiet country lanes full of clapboard farmhouses. The serenity of the mo-ment was suddenly interrupted when a pair of clowns, a little one and a big one, bolted out of a big red barn. Then an acrobat jumped off the roof. I did a double take.

We ended up chatting with the little clown—a teenage boy named Mike. He told us this was the home of Circus Smirkus, the only traveling youth circus in the US. Teenagers come to Greens-boro from all over the world to train and then perform. We stood and watched the acrobats with something close to wonder. It was an amazing sight to see—people leaping and flying through the air in total defiance of gravity—in this remote corner of Vermont. One of the acrobats, an instructor, had white hair. Another yelled in Russian. Mike explained this was his second summer with Smirkus, and each summer had its own theme. This summer, it was *The Invention of Flight*. He told me that in addition to the older persons on the circus staff, older people support the circus in a big way by attending shows, night after night, and by heavily donating, which allows the circus to offer scholarships to those who couldn't afford to attend otherwise.

We kept walking and noticed banners and fliers for a summer chamber music series pinned to garage and shed doors. In a week, a string quartet was coming from New York, and at the end of the month, an older cellist was flying in from Los Angeles. A man feeding chickens in his yard saw us squinting at the flier pinned to his shed and waved. He walked over to ask if we were Summer People.

"We're Afternoon People, actually," I told him. "Just passing through." I added that we had sort of stumbled on the town, and that I was intrigued, given the topic of my book, since it seemed to be a haven for aging people. Harold smiled, nodded his head, and said he would be happy to give examples of how this is the case. First, though, he asked my daughters if they wanted to feed his hens.

While my daughters tossed grain around the coop, I chatted with eighty-one-year-old Harold Gray over some homemade lemonade (everyone in Greensboro, it seems, has a thirst for the stuff). He had spent a few years in Cameroon with the Peace Corps, built a subsequent career at the US Agency for International Development based in D.C., and retired a few years ago to Greensboro, where he found a new vocation writing articles and shooting photos for the local newspaper. What he loves about the place is its incredible sense of community and the many opportunities for older persons to become involved in meaningful activities.

It turns out, Harold belongs to two social groups modeled on the fabled ROMEOs. The first meets for breakfast each week to discuss politics with current local office holders and political hopefuls. (These incumbents and political aspirants also meet with the Greensboro Ladies' Walking Society.) Harold's other group meets weekly to have lunch and muse about the big ideas in life. Recent topics include the future of interstellar space travel (the group includes a retired astronaut and a planetary scientist) and questions such as "Is sarcasm genetic?"

Harold showed me a picture of the lunch group: a bunch of smiling men sitting around a table. He started identifying them: "That's a civil engineer; that guy used to be a minister; that one's a retired science teacher; that's a professor who writes about the American Revolution." He noted that one of them, Tim, operates the town's only garage and is regularly voted the moderator at the Greensboro town meetings. "He's able to bring in voices of all the generations in a thoughtful way. He has that kind of respected village acceptance."

I found it inspiring that Harold's two groups were organized to promote "friendship"—a word he came to again and again in describing them. For in the US, older men are more likely to suffer from social isolation and loneliness than younger men or older women.[2] In contrast, Harold described a community in which links seemed to continue or even grow among his peers.

Harold asked my husband and me if we had read *Crossing to Safety*, by Wallace Stegner. We hadn't. He told us the book is about Stegner escaping to Greensboro every summer and the title is based on a poem by Robert Frost, which refers to "crossing to a better place, a more serene and more emotionally stable place."

He then shared his theory about Greensboro, whose population of seven hundred swells to over two thousand in the summers. "People spend their summers here, and then sometimes they'll retire here, which makes Greensboro different from a lot of other places in that its residents' extended families often live in other parts of the country. And so people depend more than usual on their friendships and build a kind of second extended family from their friends and neighbors in Greensboro."

That might help to explain why Harold is active in politics and journalism and delivers free meals to older people in need and is active on the board of trustees at the public library. His wife tutors local kids, helps out at the Nature Conservancy, and is a proud member of the Greensboro Ladies' Walking Society. I remembered what Nancy Hill, the older woman I'd met at Willey's, had told me: older people in Greensboro are *involved*.

Harold told us it would be a shame to leave town before taking a dip in Caspian Lake. So we went for a swim before hitting the road. As we were drying off on the beach in the warm July sun, I started chatting with another local, a young woman named Kathryn Lovinsky, also out for an afternoon swim with her two-year-old twin boys.

Kathryn, it turns out, directs Rural ARTS (where Carol Fairbank works). She grew up nearby and moved back with her family after living in D.C., where she went to college and lived for a few years afterward. She didn't seem particularly keen on D.C.: "Everyone was pretty isolated from one another. They'd spend the summer alone watching TV in their air-conditioned apartments. In Greensboro, people spend the summer outdoors, with each other." She told me her septuagenarian parents manage an affordable apartment complex in town, which they rent out to older people with limited savings, as well as own and operate a local business that makes sustainable packaging material. On top of that, they run an eighty-acre farm with cows and goats and chickens. And her mom regularly babysits her sons, which is a treat for everyone involved, "including me," she said with a laugh, "so I can get some time off!" She said this was pretty typical of the area: older people stay busy, raising animals or growing vegetables, and spend a lot of time engaging with the community and interacting with children and young adults.

There's clearly something special about the place. That's why people keep moving here. And what a lot of these transplants have in common is that they seem purposeful about how they want to live and grow old.

"Take Judy," Kathryn said, pointing to an older woman who had just spread her towel at the other end of the beach and was settling down with a newspaper and a bottle of sparkling water. Judy was an active real-estate agent in her eighties. She planted vegetables and exhibited her paintings at community shows every year. She was also, Kathryn informed me, a pillar of the community, involved with more committees and organizations than she could name. She started listing a few when Judy, sensing that she was the subject of discussion, looked in our direction and waved. Kathryn waved back.

We walked over. Kathryn told Judy that I was curious about Greensboro. "Well, Greensboro is the place to be," she said, like a true realtor. I told her it was thrilling to hear how involved the older population was in the governance and community life of the town. That morning, I had read dreadful stories about how both the state of New York and Poland were forcing older judges off the bench because of their age, the opposite of the ethos in Greensboro.

When I asked Judy what it is like for the older residents of Greensboro, she mentioned the infrastructure and programs that support the positive age beliefs that thrive throughout town. "It's a case of the chicken or the egg," Kathryn suggested. "Hard to say which came first, the positive age beliefs or the age-positive culture."

Judy, sitting on her big blue beach towel, gave me a little smile as she shielded her eyes from the sun. "Either way, I love it here," she said. "And I think you do too." She added that there were a few lovely houses on the market if I was interested.

I have to confess, I thought about it for a minute.

Kathryn was right when she applied the chicken-or-egg metaphor to the question of whether it was the positive age beliefs or the age-liberated culture of Greensboro that was the cause or the effect. It is likely they arose in tandem and they have been mutually supportive.

What I saw there is a vivid demonstration of what happens when older people and their surrounding society are harmonized in a productive way. Wallace Stegner, at the age of eighty-one, described this harmonization in his foreword to Greensboro's history. "I grew up without history and without the sense of belonging to anything, so desocialized that I didn't know, and don't yet, the first names of three of my grandparents. . . . Whereas Greensboro had what I lacked and wanted: permanence, tranquility, traditional and customary acceptances, a neighborly social order."[3]

Greensboro can serve as a goal for older persons who don't have the good fortune to live there. The goal would be to initiate their personal age liberation, which would contribute to the liberation of the surrounding ageist society so it, in turn, promotes the age liberation of others within it.

Greensboro illustrates another truth that was revealed to me in the process of writing this book. As a scientist, I used to assume the best way to understand the world was through an elegant graph or powerful statistical test. But in Greensboro, as in the months preceding my visit, when I met and interviewed so many inspiring older people, it became clear to me that while science helps us to discover how the world operates, stories are how we make sense of the world. "The human species," wrote the anthropologist Mary Catherine Bateson, "thinks in metaphors and learns through stories."

The stories I have learned from while writing this book include those of the residents of Greensboro, as well as the lives of Kane Tanaka and other supercentenarians who demonstrate the benefits of living in societies built on positive age beliefs. Likewise, the people in this book show how it is possible to rise above the ageism that permeates many societies. Take ninety-nine-year-old bookstore manager Irene Trenholme, who inherited her positive age beliefs from her grandmother and has found ways to challenge ageism wherever she encounters it, or Jonas, the midwestern doctor, whose negative age stereotypes fell away in later life after he spent time with older medical mentors dedicated to improving their communities. And Barbara, whose positive age beliefs were strengthened by our intervention and, as a result, found her balance and physical function greatly improved.

Whether they acquired their positive age beliefs early or later in life, many of these people built happy, healthy, successful older lives on the foundation of these beliefs. Take John Basinger, the actor, and Patrick Hamilton, the mushroom hunter: both draw on positive age beliefs to strengthen their memory. Also, Sister Madonna Buder, the triathlete, and Maurine Kornfeld, the swimmer, whose impressive athletic feats are fueled by the way they approach aging. Or Mel Brooks and Liz Lerman, whose positive age beliefs have unleashed a flurry of creativity in later life.

The potential for anyone to have a life shaped by positive age beliefs, with all the benefits it entails, was revealed to me in an unexpected way. As I was creating the Image of Aging questionnaire, to assess age stereotypes, I asked participants for the first five words that came to mind when they thought of aging. The responses were almost all negative. But when I asked for the first five words that came to mind when they thought of a positive image of aging, everybody could come up with positive words.

The discovery was unexpected because I had previously found ample evidence that older individuals internalize negative age stereotypes through exposure to the myriad institutions of society that disseminate them and through experiencing discrimination based on these stereotypes. Yet, the first-five-word question exposed the positive age beliefs that had been there all along, ready to be activated, which is the aim of the ABC method and exercises presented in Chapter 9 and Appendix 1 of this book.

The process of activating positive age beliefs has a parallel in one of my favorite poems written by Derek Walcott, the Caribbean poet who won the Nobel Prize for Literature. This poem, which he called "Love After Love," is not explicitly about aging or age beliefs, but to me it brings to life some of the central ideas explored in this book.

> *The time will come*
> *when, with elation,*
> *you will greet yourself arriving*
> *at your own door, in your own mirror,*
> *and each will smile at the other's welcome,*
>
> *and say, sit here. Eat.*
> *You will love again the stranger who was your self.*
> *Give wine. Give bread. Give back your heart*
> *to itself, to the stranger who has loved you*
>
> *all your life, whom you ignored*
> *for another, who knows you by heart.*
> *Take down the love letters from the bookshelf,*
>
> *the photographs, the desperate notes,*
> *peel your own image from the mirror.*
> *Sit. Feast on your life.*

It seems Walcott describes a change that can take place at two levels. At the individual level, he describes a metaphorical stranger, which could represent the older self that holds positive age stereotypes that have been eclipsed by negative age stereotypes. At a point when these dormant positive age stereotypes become ascendant, what had been the stranger is welcomed back.

At the societal level, Walcott's poem could be seen as a call for solidarity. It is time to remove age-based barriers and prejudices, to "give back your heart to itself, to the stranger who has loved you all your life." When older persons are no longer treated by society as strangers but are instead valued by themselves and their communities, aging can become a homecoming, a rediscovery, a feast of life.

APPENDIX 1

ABC Method to Bolster Positive Age Beliefs

	Exercising Your ABCs: The Health-Promoting Age Belief Tools
A	**Awareness:** Identifying where negative and positive images of aging are found in society
B	**Blame:** Understanding that health and memory problems can be the result, at least in part, of the negative age beliefs we aquire from society
C	**Challenge:** Taking action against ageism so that it is no longer harmful

Most of the following exercises can be learned and carried out rapidly. Since age beliefs are multifaceted, operating at both unconscious and conscious levels, it would be helpful to try a combination of these exercises, with at least one from each stage. As discussed in Chapter 9, the three stages consist of: increasing **Awareness**, placing **Blame** where blame is due, and **Challenging** negative age beliefs.

To strengthen these beliefs and to become more comfortable with them, I suggest repeating the exercises you select. What Aristotle discovered twenty-four centuries ago is still true today: "We are what we repeatedly do." Consistently applying these strategies should lead to a compounding of benefits, a snowballing effect in which small changes lead to a cascade of improvements.[1]

Here are the ABC exercises for you to try:

Awareness Exercises

Awareness Exercise 1: Five Images of Aging

Jot down the first five words or phrases that come to mind when you think of an older person. Even if you already did this exercise in Chapter 1, try it again to see whether your age beliefs have shifted since you started reading this book. Again, there are no right or wrong answers. How many of your responses are negative and how many positive?

If you find yourself with lots of negatives in the Five Images exercise, that doesn't mean your views are set in stone. Most of us have unconsciously assimilated negative age beliefs from our surroundings, but we can reverse these beliefs. Becoming aware of them is the first step.

Awareness Exercise 2: Portfolio of Positive Role Models

Who are your older role models? List four older people you admire. Pick a couple from your own life and others from the world at large, such as history, books (including this one), TV shows, or current events. In that way, you'll collect a diverse set of role models and associate a range of admirable qualities with aging. For each model, pick one or more qualities you admire and would like to strengthen in yourself as you get older.

Awareness Exercise 3: Noticing Age Beliefs in Media

A good way to make visible the invisible is to record both negative and positive images of aging that you encounter in the course of just one week, using a notebook or your smartphone. When you watch TV or stream shows, take note of whether there are any older characters, what roles they play, and whether these paint aging in a negative or positive light. As you spend time online or read the newspaper, write down how older persons are included and note when they aren't included. At the end of the week, tally up the number of negative and positive images of aging, as well as the number of omissions. In my studies, I found that this kind of active noticing helps develop a keen awareness of not just blatant ageism, but also the more subtle forms of exclusion and marginalization.[2]

Awareness Exercise 4: Awareness of Generations

Think about your five closest friends. If you're like me, these five people probably have birthdays within a couple of years of yours. Of course, there's nothing wrong with enjoying the company of your age peers, but the ease with which we keep strictly to ourselves, generationally speaking, is another enabler of negative age beliefs. Think about how to increase your intergenerational contact. Take a look at how many meaningful intergenerational interactions you had in the last week. If you have trouble thinking of many, come up with two activities you could undertake in the next month that involve different generations.

Blame-Shifting Exercises

Blame-Shifting Exercise 1: Find the Real Cause

Monitor yourself for when age stereotypes influence how you think about the cause of unpleasant events or challenges. If you or an older person you know loses keys or forgets a date or name, and you find yourself leaning on the term "senior moment," remember that this is your negative age belief speaking, rather than an objective assessment of the aging process. Is it possible that you or the other person was rushed, stressed, saddened, or distracted by something when the information was being encoded or retrieved? Those emotional states can all heighten temporary forgetfulness. If you blame a sore back or not hearing something on aging, notice the circumstances: Did you pick up something that was too heavy or is the background noise too loud? Think of two actual or hypothetical mental or physical incidents that happened to you or another older person and was blamed on aging. Then, think of a cause that has nothing to do with aging to explain the incident.

Blame-Shifting Exercise 2: Who Profits?
Write down four negative age stereotypes. Name a company or institution that might benefit or profit from such a stereotype. For example, if you wrote down "memory loss," you might then list Lumosity, a company that sells "brain training games," often by drawing on the anxiety associated with the negative age belief that all memory inevitably declines. The company was sued by the Federal Trade Commission for preying on older consumers' fears with false statements.[3]

Blame-Shifting Exercise 3: Sexist If It Were About a Woman?
If you aren't sure whether a reference to or action against older people is ageist, try switching it so that the target is another marginalized group, such as women. For example, if an employer states the need to fire older workers, ask yourself how it would sound if the same comment was made about firing women. If it would sound sexist, consider labeling it ageist when older workers are targeted.

Challenge Exercises

Challenge Exercise 1: Dismantling Negative Age Beliefs
You can challenge negative age beliefs by presenting accurate information. This book covers much of the science that disproves common negative age stereotypes. (It is summarized in Appendix 2: Ammunition to Debunk Negative Age Stereotypes.) Write down

three myths about aging. Practice what you might say to someone who thinks they are true. For instance, "The old don't care about the planet." Turns out those over sixty-five recycle more than any other age group (and recycling rates go up as people get older).[4]

If you're like me, you may not always have a zippy one-liner available right when someone says something ageist. So it helps to have a few lines at the ready, or to circle back to make a comment at a later time that challenges an earlier ageist comment or action.

Challenge Exercise 2: Find Ways to Get Involved in Politics

You can run for political office. Alternatively, determine which candidates have advocated for public policies that contribute to the well-being of older constituents and then support their campaigns for office. You could also let your elected representatives know when you agree or disagree with their positions on legislation that is relevant to older constituents.

Challenge Exercise 3: Confront Media Ageism

When you read an article that reflects negative age stereotypes, write to the editor or post about it on social media. One recent example is an ad that E-Trade (the online stock-trading platform) unveiled at the 2018 Super Bowl, the most watched sporting event in the US. The ad ridicules older people for working: we see an older mail carrier dropping a stack of packages; an older fireman being lifted off the ground when he points the hose at the sidewalk instead of at the fire. The elderly dentist and sports ref are no less bumbling and incompetent. Just in case the older age of the workers is not sufficiently emphasized, the words to the song playing in the background are "I'm 85 and I Want to Go Home," sung to the tune of Harry Belafonte's ballad "Day-O."

The advertising agency apparently created the ad to scare younger potential customers into spending for early retirement by giving commissions to E-Trade for trading stocks.[5] Even though these negative images helped produce a large profit for E-Trade over the following year,[6] the ad also generated anger and backlash. And there, by the way, is upstream causality in a nutshell: ageism driven not by facts, but by good old-fashioned hunger for profit.

The first time I heard about this ad was through one of my daughters, who caught wind of it on her Facebook feed and showed me post after post from friends and strangers alike expressing disgust at the way the ad portrayed older people.

Keep your eye out for the next ageist example to appear and find a way to register your concern, by sending a protest message to the company whose product is advertised or by organizing a petition to let the company know that if it continues you and your friends will do business with age-friendlier companies.

APPENDIX 2

Ammunition to Debunk Negative Age Stereotypes

The following presents examples of the false and harmful age stereotypes that are disseminated by a wide range of societal sources. These stereotypes are accompanied by a selection of the considerable evidence that you might find helpful in refuting them (with references in the endnotes).

1. **FALSE AGE STEREOTYPE**: The saying "You can't teach an old dog new tricks" applies to older people's inability to learn.
   FACT: There are many positive cognitive changes in older age and there are many techniques to support lifelong learning. Older persons can benefit from the same memory strategies that young persons use to improve recall. In fact, our brains experience new growth of neurons in response to challenges throughout the life span.[1,2, 3,4]
2. **FALSE AGE STEREOTYPE**: All older persons experience dementia.
   FACT: Dementia is not a normal part of aging. Most older persons do not experience dementia. Only about 3.6 percent of US adults aged sixty-five to seventy-five have dementia. Further, there is evidence that dementia rates have been declining over time.[5,6,7]
3. **FALSE AGE STEREOTYPE**: Older persons' health is entirely determined by biology.

FACT: Our team has found that culture, in the form of age beliefs, can have a powerful influence on the health of older persons. For instance, positive age beliefs can benefit their health in multiple ways, such as reducing cardiovascular stress and improving memory. In contrast, negative age beliefs can have a detrimental impact on these aspects of health.[8,9,10,11] We also found that positive age beliefs amplified the beneficial impact of *APOE* ε2, a gene that often benefits cognition in later life.[12]

4. **FALSE AGE STEREOTYPE**: Older persons are fragile, so they should avoid exercise.

   FACT: Most older persons can exercise without injury. The World Health Organization recommends that older persons regularly exercise because this can benefit cardiovascular and mental health, as well as lead to stronger bones and muscles.[13]

5. **FALSE AGE STEREOTYPE**: Most older people suffer from mental illness that can't be treated.

   FACT: Most older persons do not suffer from mental illness. Studies show that often happiness increases, whereas depression, anxiety, and substance abuse decline in later life.[14] Further, older persons usually benefit from mental health treatment including psychotherapy.[15,16]

6. **FALSE AGE STEREOTYPE**: Older workers aren't effective in the workplace.

   FACT: Older workers take fewer days off for sickness, benefit from experience, have strong work ethics, and are often innovative.[17,18,19] Teams that include older persons have been found to be more effective than teams that do not.[20]

7. **FALSE AGE STEREOTYPE**: Older persons are selfish and don't contribute to society.

   FACT: Older persons often work or volunteer in positions that allow them to make meaningful contributions to society.

They are the age group that is most likely to recycle and make philanthropic gifts. In older age, altruistic motivations become stronger, while narcissistic values wane in influence. Older persons often engage in legacy thinking, which involves wanting to create a better world for future generations. Also, in most families, there is a downward flow of income with more funds going from older adults to grown children than from grown children to older adults.[21,22,23,24,25,26]

8. **FALSE AGE STEREOTYPE**: Cognition inevitably declines in old age.

   FACT: A number of types of cognition improve in later life, among them: metacognition or thinking about thinking; taking into account multiple perspectives; solving interpersonal and intergroup conflicts; and semantic memory. Other types of cognition tend to stay the same, such as procedural memory, which includes routine behaviors like riding a bike.[27,28,29,30] Further, I have found that strengthening positive age beliefs can successfully improve the types of memory that are thought to decline in later life.[31,32,33,34,35]

9. **FALSE AGE STEREOTYPE**: Older persons are bad drivers.

   FACT: The absolute number of crashes involving older drivers is low. They are more likely to use seat belts and follow speed limits. Also, they are less likely to drive while texting, while intoxicated, or at night.[36,37,38]

10. **FALSE AGE STEREOTYPE**: Older persons don't have sex.

    FACT: Most older persons continue to enjoy a physically and emotionally fulfilling sex life. A survey found that 72 percent of older adults have a romantic partner and, of those, most are sexually active.[39,40]

11. **FALSE AGE STEREOTYPE**: Older persons lack creativity.

    FACT: Creativity often continues and even increases in later life. Numerous artists, including Henri Matisse, are credited with producing their most innovative works at an older age. Successful start-ups are more likely to be run by entrepreneurs over fifty than under thirty. Older persons are often leaders in innovation and use it to revitalize communities.[41,42,43,44]

12. **FALSE AGE STEREOTYPE**: Older persons are technologically challenged.

FACT: Older persons possess the ability to adapt to, learn, and invent new technology. Three-quarters of those fifty and older use social media on a regular basis; 67 percent of those sixty-five and older use the internet and 81 percent aged sixty to sixty-nine use smartphones.[45,46] Some older persons have led advances in technology, including MIT professor Mildred Dresselhaus, who innovated the field of nanotechnology in her seventies.[47]

13. **FALSE AGE STEREOTYPE**: Older persons don't benefit from healthy behaviors.

FACT: It is never too late to benefit from healthy behaviors. For example, older persons who quit smoking show improvement to their lung health within a few months.[48] Similarly, older persons who overcome obesity show improvement to cardiovascular health.[49]

14. **FALSE AGE STEREOTYPE**: Older persons don't recover from injury.

FACT: Most older people who become injured show recovery, and older persons with positive age beliefs are significantly more likely to fully recover.[50]

A Call to End Structural Ageism

The miracle of longevity provides such incredible opportunity to individuals and the societies in which we live. Yet, today so much of that potential remains unrealized because we haven't adequately addressed these challenges that hinder older populations from living their later lives in meaningful, productive ways.
—*Paul Irving, Chair, Milken Institute Center for the Future of Aging*[1]

The best way to eliminate negative age stereotypes is to end structural ageism. Since this ageism is deeply rooted in the power structure of society, to achieve social change requires multifaceted activities from two directions: top-down, which would involve laws and policies; and bottom-up, which would involve an age liberation movement demanding these changes. The following is a partial list of what is needed to achieve age justice. I encourage you to consider whether there is a sector or an item that you could impact.

End Ageism in Medicine

- End age discrimination in providing treatments for a range of illnesses, including cardiovascular disease and cancer. In 85 percent of 149 studies, health-care providers excluded older patients from, or were less likely to offer them, procedures and treatments even when they were equally likely to benefit as younger patients.[2]

- Increase support of preventive care and rehabilitation services for older persons[3] through better health insurance reimbursement.

- Improve how health-care providers communicate with their older patients. This would include avoiding patronizing language and ending the practice of excluding older patients from important health-care decisions. To improve current practice, geriatrician Mary Tinetti developed an effective conversation guide to help health-care providers take into account the priorities of older patients.[4]

- Create geriatric emergency departments at all hospitals. In the US, hospitals often have pediatric emergency departments, but only 2 percent have geriatric emergency departments; these have resulted in improved health care for older persons and reduced costs.[5]

- End the salary and reimbursement disparity among health-care professionals by which those who focus on older persons are paid less than those of other medical specialties.[6]

- Expand the number of departments of geriatrics so that they are included in all medical schools. Of the 145 medical schools in the US, only 5 have departments of geriatrics; there is about one geriatrician for every three thousand older Americans.[7]

- Provide geriatric training for all health-care providers so they are prepared to care for older patients. Training could include diverse older patients with a range of health levels. In the US, while all medical schools require pediatric training, less than 10 percent require geriatric training.[8] Similarly, less than 1 percent of nurses and less than 2 percent of physical therapists are formally trained to work with older adults.[9]

- Include antiageism content in the training of health-care professionals, which could include overcoming the widely held myth that, for instance, hypertension and back pain are inevitable in later life.[10]

- Include age-belief screening in primary care visits for all patients and prescribe strategies to challenge negative age beliefs.
- Overcome age discrimination in providing proper screening and referrals for mental health issues, STDs, and elder abuse by putting in place standard protocols for older patients and training for health-care providers to carry out the protocols.[11]

End Ageism in Mental Health Care

- Reform mental health training so that it adequately includes issues related to older persons, such as findings that depression is not a natural part of aging and that older persons often have skills that allow them to benefit from psychotherapy.
- End the practice by which Medicare reimburses therapists who treat older patients at rates that are substantially lower than the market rates.[12]
- Add information about the mental health of older persons to the *Psychodynamic Diagnostic Manual* and the *Diagnostic and Statistical Manual of Mental Disorders*, which are used as guides by mental health professionals.
- Establish intergenerational psychotherapy groups so people of different ages can learn from one another.
- Reduce the gap between mental health need and care that increases with older age in many countries. This can be done by increasing treatment options by, for example, expanding the Friendship Bench model of older persons administering lay mental health care beyond the countries where it currently operates.[13]

End Ageism in the Governmental System

- Establish and enforce legislation that provides older persons with economic and food security. In the US, 9 percent of older persons live in poverty, 16 percent of older persons do not have adequate food, and 306,000 are homeless.[14]
- Create an antiageism tsar and an antiageism agency on the federal level to initiate and coordinate antiageism policies across all government departments.
- Encourage older persons to run for political office at all levels in order to advocate for age-friendly policies, and to become involved in political campaigns of those who support their interests.
- Include the protection of older persons' rights in all laws relating to civil rights. Many such laws, including the US Civil Rights Act, do not include age.[15]
- Improve conditions in nursing home and long-term-care facilities through laws that require adequate staffing levels, training, and compensation.
- Prohibit nursing homes and long-term care facilities from inappropriately using medications to sedate older residents. According to several recent reports, a number of American nursing homes use sedating medications to manage dementia symptoms, even though the Food and Drug Administration never approved many of them for this use, and they can cause fatigue, falls, and cognitive impairment.[16]
- Provide funding for law enforcement and programs designed to prevent and stop elder abuse, which social epidemiologist E-Shien Chang found is determined by modifiable factors.[17,18]
- Make voting easily accessible to all older people, by such means as providing transportation to polling stations and by making absentee ballots readily accessible.
- Demand that all countries ratify the UN Convention to

Strengthen the Human Rights of Older Persons. Multiple countries, including the US, have failed to do so.[19]

- Ensure that older persons are adequately included in juries and on judicial benches. A lack of inclusion of older persons in these roles has become a growing problem.[20]

End Ageism in Education

- Advocate that school boards develop goals for preschool to grade twelve curricula that include positive depictions of older people in courses, such as history and social studies. Many curricula now include other diversity goals, but do not include diversity by age as an aim.
- Encourage teachers to include positive portrayals of older persons through films, songs, events, and books in their classes. For example, see the relevant list of children's literature developed by education activist Sandra McGuire.[21]
- Expand college and graduate school courses in developmental psychology, most of which do not go beyond young adulthood, to include aging topics.
- Include ageism awareness in teacher training, which would show how ageist messages are transmitted in schools and how they can be countered.
- Support programs that bring older persons from the community into schools to talk about what they have accomplished and to engage in mentorship opportunities. The latter was initiated by Columbia School of Public Health dean Linda Fried's Experience Corps; it should be expanded to all schools.[22]
- Set up "Grandparents Day" during which students celebrate older relatives or older people from the community.
- Enhance educational opportunities for older persons, ranging

from literacy programs for those who did not have educational opportunities early in life to making courses available in universities and colleges. The Age Friendly University initiative, which promotes intergenerational learning, could be expanded to the 98 percent of universities in the world that have not adopted its age-inclusive principles.[23]

End Ageism in the Workplace

- End ageism in hiring older workers by adequately enforcing anti-age-discrimination laws.
- End the firing of workers based on age, including forced retirement. For example, United Nations employees, including those that focus on issues of aging, are forced to retire by the age of sixty-five.[24]
- Incorporate older age in diversity, equity, and inclusion training programs and policies. This could raise awareness about the age discrimination that is currently reported by 60 percent of workers, dispel myths about older workers, and highlight the contributions of older workers. A survey of employers in seventy-seven countries found that only 8 percent include age in diversity, equity, and inclusion policies.[25]
- Set up whistleblowing programs in which older workers who leave jobs or retire could share with the public any experiences of ageism, without risk of punishment by employers.
- Implement intergenerational work teams when possible. These teams have been found to shatter age stereotypes and increase productivity.[26]
- Establish a system that rates companies on how age positive they are and award certificates to those that are the most age friendly.

End Ageism in the Antiaging
and Advertisement Industries

- Monitor the negative age stereotypes that companies present in advertisements; this could include using an online clearinghouse to which individuals would provide examples of ageist ads.
- Organize a boycott of companies that demean older persons in their advertisements, including the many created by the antiaging industry, until they agree to end the offensive messages.
- Increase the inclusion and diversity of older persons in advertisements and present them in roles that reflect vitality. To challenge the negative and stereotypical presentation of older persons, the Centre for Ageing Better in the UK recently launched the first online archive of positive and realistic images of older people that are freely available.[27]
- Give older people a seat at the table as creative directors at advertising agencies. The average age of advertising agency employees is thirty-eight, even though most consumers are over the age of fifty.[28]
- Establish awards for ads that empower older persons.

End Ageism in Popular Culture

- Expand the meaning of "diversity" in films to include older actors, writers, and directors. The Academy of Motion Picture Arts and Sciences, which presents the Academy Awards, excludes older persons in its new diversity inclusion rules.[29]
- Monitor and publicize the rampant ageism in movies and television, which includes both ageist language and activities, as well as the absence of nuanced older characters.[30] Let producers and fellow viewers know that this is not acceptable.

- Recruit and support celebrities to speak out against ageism in Hollywood and in the larger culture. A number of celebrities such as Amy Schumer, Madonna, and Robert De Niro have spoken out.[31] More voices are needed.
- Create a national holiday celebrating older persons with local events. Japan has a national holiday of older persons that can serve as a model.
- Organize some of the billions of gamers to boycott current video games that include ageist content,[32] and encourage the video-game industry to produce games with positive portrayals of aging.
- Foster the creation and sale of age-positive birthday cards. These could replace the ubiquitous ones that denigrate aging. Efforts of local artists and activists to take on these commercial markers of aging have begun in Colorado and the UK.[33]
- Create 50 Over 50 campaigns. This could be modeled after the 30 Under 30 age-based lists that are publicized by various industries to recognize leaders in their fields.

End Ageism in Media

- Pressure the government to prohibit digital age discrimination that excludes older persons from housing and job listings. Under the current system, social media companies are supposed to self-police, but this has not been successful.[34]
- Demand that social media companies ban the ageism it disseminates. Facebook's community standards should prohibit ageism in the same way it bans hate speech directed at other groups, and Twitter should enforce its community standards that prohibit ageism.[35,36] Evidence that it is not yet being enforced comes from a Twitter analysis that found 15 percent of the tweets under the #BoomerRemover hashtag were characterized by overt de-

rogatory statements that included wishing death upon the older generation.[37]

- Encourage journalism schools to stress the importance of reporting about structural ageism as well as writing news stories that empower older persons.[38] A model that can be used for this is the Age Boom Academy run by Columbia School of Journalism and the Robert N. Butler Columbia Aging Center of the Columbia School of Public Health.
- Replace the use of ageist language and concepts in news stories, such as the use of the term "silver tsunami" to describe the aging of the Baby Boomer generation, with alternatives, such as "silver reservoir," which reflects the idea that this generation can be a "potential resource for good in our society rather than an impending danger that threatens to wipe everyone out."[39]
- Ask media outlets to provide time and space on TV and radio and in newspapers for interests of their older viewers and readers. *The New York Times* reporter Paula Span's column "The New Old Age" is a model for this.
- Create journalism prizes for outstanding antiageism and pro-aging reporting.

End Spatial Ageism

- Eliminate the age-based digital divide by which older persons are significantly less likely to have access to the internet at home than younger persons. This lack of internet access, which now impacts 42 percent of Americans who are sixty-five or older, is particularly severe for older persons who are low income, female, living alone, immigrants, disabled, and members of ethnic minority groups.[40] As connecting online can facilitate health care, work opportunities, and community involvement, it is imperative that governments provide affordable connectivity with adequate technical support for all older persons.

- End zoning regulations and regional planning that segregates and isolates housing for older persons.
- Reduce the social isolation of older persons by insisting that the government provide adequate age-inclusive and accessible public transportation in both urban and rural areas.[41]
- Require that housing complexes built with federal funding include older persons in numbers that are at least as high as their proportion of the general population.
- Facilitate in-person intergenerational contact through such means as designing age-inclusive public and private spaces including libraries, museums, and multiuse parks.
- End the neglect of older people during natural disasters that can leave them trapped in dangerous places[42] by including them on an equitable basis in natural disaster emergency-relief plans.

End Ageism in Science

- End the practice of excluding older persons from clinical trials, which even happens when they are particularly likely to have the targeted illnesses, such as Parkinson's disease.[43] Their inclusion should be required at a rate that is at least proportionate to the total population, to ensure medications and treatments are safe and efficacious for older individuals.
- Create surveys that include older participants and that report whether and, if so, how older persons are resilient and how they experience illness, treatments, and recovery differently than other age groups. Most surveys do not collect data for those over age sixty-five;[44] exceptions are the Health and Retirement Study and

its sister studies, the Baltimore Longitudinal Study of Aging and the UK Biobank.

- End the use of the term "dependency ratio" that is frequently used in science and policy reports; it characterizes everyone in a population who is sixty-five or older as being dependent on younger adults and not productive members of society.
- Increase funding for aging research that includes the biological, psychological, and societal determinants of aging health, as well as studies of the best policies and programs to take advantage of increased longevity. Less than 0.01 percent of the US federal budget is devoted to aging research, and less than 1 percent of US foundation funding is devoted to it.[45]
- Change the common definition of aging as "senescence," a process of progressive decay, to a more multidisciplinary and positive definition, such as a later-life developmental stage that can include psychological, biological, and social growth based on decades of accumulated experience.

ACKNOWLEDGMENTS

I would like to acknowledge the following people who contributed to the book in important ways through their stories, knowledge, and inspiration. Many generously gave hours of their time to chat. These include:

Carl Bernstein, John Blanton, Bethany Brown, Madonna Buder, Robert Butler, Jennifer Carlo, Neil Charness, Dixon Chibanda, Kinneret Chiel, Jessica Coulson, Wilhelmina Delco, Thomas Dwyer, Carol Fairbank, Rachella Ferst, Judy Gaeth, Susan Gianinno, Stacey Gordon, Harold Gray, Angela Gutchess, Patrick Hamilton, Nancy Hill, Paul Irving, Maurine Kornfeld, Nina Kraus, Suzanne Kunkel, Jack Kupferman, Liz Lerman, Vladimir Liberman, Kathryn Lovinsky, Richard Marottoli, Deborah Miranda, Piano Noda, Helga Noice, Tony Noice, Daniel Plotkin, David Provolo, Nancy Riege, Elisha Schaefer, Bridget Sleap, David Smith, Wilhelmina Smith, Quinn Stephenson, Jemma Stovell, Kane Tanaka, Irene Trenholme, Christopher Van Dyck, Yumi Yamamoto, Robert Young, the Anti-Amyloid Treatment in Asymptomatic Alzheimer's Disease participants, and the Wrinkle Salon participants.

It takes a village to conduct research. I could not have carried out the research described in these pages without the expertise of many people. I feel lucky to have been able to work with the following excellent colleagues: Heather Allore, Kimberly Alvarez, Ori Ashman, Mahzarin Banaji, Avni Bavishi, Eugene Caracciolo, E-Shien Chang, Pil Chung, Mayur Desai, Lu Ding, Margie Donlon, Theodore Dreier, Itiel Dror, Thomas Gill, Jeffrey Hausdorff, Rebecca Hencke, Sneha Kannoth, Stanislav Kasl, Julie Kosteas, Suzanne Kunkel, Rachel Lampert, Ellen Langer, Deepak Lakra, John Lee, Erica Leifheit-Limson, Sue Levkoff, Samantha Levy, Sarah Lowe, Richard Marottoli, Jeanine May, Scott Moffat, Joan Monin,

Terry Murphy, Lindsey M. Myers, Kristina Navrazhina, Reuben Ng, Linda Niccolai, Robert Pietrzak, Corey Pilver, Natalia Provolo, Kathryn Remmes, Susan Resnick, Mark Schlesinger, Emma Smith, Mark Trentalange, Juan Troncoso, Sumiko Tsuhako, Peter Van Ness, Shi-Yi Wang, Jeanne Wei, and Alan Zonderman. I am especially indebted to the biostatistical brilliance of Marty Slade, and the epidemiological brilliance of the National Institute on Aging's scientific director, Luigi Ferrucci.

I also would like to thank the individuals who participated in our studies, and the investigators who conducted the longitudinal studies that I have been able to draw on: the Baltimore Longitudinal Study of Aging, the Ohio Longitudinal Study on Aging and Retirement, the Health and Retirement Study, the Precipitating Events Project, and the National Health and Resilience in Veterans Study.

Much of the research described would not have been possible without the generous financial support of the National Institute on Aging, the Patrick and Catherine Weldon Donaghue Medical Research Foundation, the National Science Foundation, the Yale Program on Aging, and the Brookdale Foundation.

There are many people who I would like to thank for helping to create this book. First, I am grateful to Elissa Epel, who suggested the time is right to disseminate my team's findings beyond journal pages.

Thanks to colleagues, friends, and family members who carefully read drafts of chapters and offered invaluable feedback. These include: Jose Aravena, Andrew Bedford, E-Shien Chang, Benjamin Levy, Charles Levy, Elinor Levy, Samantha Levy, Lisa Link, Eileen Mydosh, and Renee Tynan. I am thankful to Natalia Provolo for her thoughtful reading and careful checking of facts and references. I also appreciate the help of Mao Shiotsu, who coordinated activities in Japan.

I am grateful for my literary agency team, creatively led by Doug Abrams, which was helpful at every stage of the book process. The team included Lara Love, Ty Love, and Jacob Albert, who provided assistance in the composition of the book and who helped me make the leap to writing for a broader audience. I also appreciate the excellent editorial advice of Rachel Neumann.

And many thanks to my editor, Mauro DiPreta, for his consistent support, enthusiasm, and insight at the many steps of turning my proposal into a book. I would also like to thank his highly capable team, which included copy editor Laurie McGee, associate editor Vedika Khanna, senior marketing director Tavia Kowalchuk, and publicity manager Alison Coolidge.

I am indebted to the Social and Behavioral Sciences Department of the Yale School of Public Health and the Psychology Department of Yale University for providing me with a collaborative environment that attracts wonderful colleagues and students. I also appreciate the support of Dean Sten Vermund, who encourages creative avenues for communicating science and provided me with a sabbatical that I could devote to writing.

I am so grateful to my family for their loving support and encouragement to get out of my comfort zone to work on this book. I feel thankful to my daughters, Talya and Shira, who have kept me aware of trends in popular culture and inspired me by showing there are many ways to fight for a just society. I also appreciate Talya's sharing a relevant text about Sigmund Freud and Shira for teaching me how optical illusions reveal how our brains work.

They say that you can't pick your parents. But if I could have, I would have picked the ones I was born to. My mom, Elinor, has inspired me to see that a woman scientist could lead a lab and balance this with raising a family and keeping a spiritual center. She advised me on the biological factors that impact health. I also am indebted to my father, Charles, who advised in almost all stages of the writing process. My dad, the best sociologist I know, has taught me the value of observing social dynamics and thinking about the causes that are sometimes hidden.

I am also thankful to my husband, Andy, who has been the ideal partner in writing this book. He has been a constant source of calmness, has offered his insights into the medical world, and has encouraged me to take on the challenge of sharing my scientific findings with a larger audience. He also knows just the right time to break out in a silly dance.

Lastly, I would like to thank you for taking the time to read this book and think about the steps needed to bring about age liberation.

NOTES

Introduction: Ideas Bouncing Between the US and Japan

1. Tsugane, S. (2020). Why has Japan become the world's most long-lived country: Insights from a food and nutrition perspective. *European Journal of Clinical Nutrition, 75,* 921–928. https://doi.org/10.1038/s41430-020-0677-5.
2. Lock, M. (1995). *Encounters with aging: Mythologies of menopause in Japan and North America.* Berkeley: University of California Press.
3. Bribiescas, R. G. (2016). *How men age: What evolution reveals about male health and mortality.* Princeton, NJ: Princeton University Press; Bribiescas, R. G. (2019). Aging men. Morse College Fellows Presentation. Yale University.
4. Levy, B. (1996). Improving memory in old age through implicit self-stereotyping. *Journal of Personality and Social Psychology, 71,* 1092–1107; Levy, B. R., Pilver, C., Chung, P. H., & Slade, M. D. (2014). Subliminal strengthening: Improving older individuals' physical function over time with an implicit-age-stereotype intervention. *Psychological Science, 25,* 2127–2135; Levy, B. R., Slade, M. D., Kunkel, S. R., & Kasl, S. V. (2002). Longevity increased by positive self-perceptions of aging. *Journal of Personality and Social Psychology, 83,* 261–270; Levy, B. R., Slade, M. D., Murphy, T. E., & Gill, T. M. (2012). Association between positive age stereotypes and recovery from disability in older persons. *JAMA, 308,* 1972–1973; Levy, B. R. (2009). Stereotype embodiment: A psychosocial approach to aging. *Current Directions in Psychological Science, 18,* 332–336.
5. Ritchie, H. (2019, May 23). The world population is changing: For the first time there are more people over 64 than children younger than 5. Our World in Data. https://ourworldindata.org/population-aged-65-outnumber-children.
6. Levy, B. R., Slade, M. D., Kunkel, S. R., & Kasl, S. V. (2002). Longevity increased by positive self-perceptions of aging. *Journal of Personality and Social Psychology, 83,* 261–270.

Chapter 1: The Pictures in Our Head

1. Levy, B. R., Slade, M., Chang, E. S., Kannoth, S., & Wang, S. H. (2020). Ageism amplifies cost and prevalence of health conditions. *The Gerontologist, 60,* 174–181.

2. Bargh, J. (2017). *Before you know it: The unconscious reasons we do what we do*. New York: Touchstone; Soon, C. S., Brass, M., Heinze, H. J., & Haynes, J. D. (2008). Unconscious determinants of free decisions in the human brain. *Nature Neuroscience, 11,* 543–545.
3. Banaji, M. R., & Greenwald, A. G. (2013). *Blindspot: Hidden biases of good people*. New York: Delacorte Press, p. 67.
4. Moss-Racusin, C. A., Dovidio, J. F., Brescoll, V. L., Graham, M. J., & Handelsman, J. (2012). Science faculty's subtle gender biases favor male students. *Proceedings of the National Academy of Sciences, 109,* 16474–16479.
5. Kang, S. K., DeCelles, K. A., Tilcsik, A., & Jun, S. (2016). Whitened résumés: Race and self-presentation in the labor market. *Administrative Science Quarterly, 61,* 469–502.
6. Bendick, M., Brown, L. E., & Wall, K. (1999). No foot in the door: An experimental study of employment discrimination against older workers. *Journal of Aging and Social Policy,10,* 5–23; Fasbender, U., & Wang, M. (2017). Negative attitudes toward older workers and hiring decisions: Testing the moderating role of decision makers' core self-evaluations. *Frontiers in Psychology.* https://doi.org/10.3389/fpsyg.2016.02057; Kaufmann, M. C., Krings, F., & Sczesny, S. (2016). Looking too old? How an older age appearance reduces chances of being hired. *British Journal of Management, 27,* 727–739.
7. Rivers, C., & Barnett, R. C. (2016, October 18). Older workers can be more reliable and productive than their younger counterparts. Vox. https://www.vox.com/2016/10/18/12427494/old-aging-high-tech; Börsch-Supan, A. (2013). Myths, scientific evidence and economic policy in an aging world. *The Journal of the Economics of Ageing, 1–2,* 3–15; Schmiedek, F., Lövdén, M., & Lindenberger, U. (2010). Hundred days of cognitive training enhance broad cognitive abilities in adulthood: Findings from the COGITO study. *Frontiers in Aging Neuroscience, 2,* 27. https://doi.org/10.3389/fnagi.2010.00027.
8. Wyman, M. F., Shiovitz-Ezra, S., & Bengel, J. (2018). Ageism in the health care system: Providers, patients, and systems. In L. Ayalon & C. Tesch-Römer (Eds.), *Contemporary perspectives on ageism* (pp. 193–212). Cham, Switzerland Springer International, 193–212; Hamel, M. B., Teno, J. M., Goldman, L., Lynn, J., Davis, R. B., Galanos, A. N., Desbiens, N., Connors, A. F., Jr., Wenger, N., & Phillips, R. S. (1999). Patient age and decisions to withhold life-sustaining treatments from seriously ill, hospitalized adults. SUPPORT Investigators. Study to understand prognoses

and preferences for outcomes and risks of treatment. *Ann Intern Med.* (1999 January 19). 130(2):116–125; Stewart, T. L., Chipperfield, J. G., Perry, R. P., & Weiner, B. (2012). Attributing illness to "old age": Consequences of a self-directed stereotype for health and mortality. *Psychology and Health, 27,* 881–897.

9. Levy, B. R. (2009). Stereotype embodiment: A psychosocial approach to aging. *Current Directions in Psychological Science, 18,* 332–336; Levy, B. R., Slade, M. D., Pietrzak, R. H., & Ferrucci, L. (2020). When culture influences genes: Positive age beliefs amplify the cognitive-aging benefit of *APOE* ε2. *The Journals of Gerontology, Series B: Psychological Sciences and Social Sciences,, 75,* e198–e203.

10. Chang, E., Kannoth, S., Levy, S., Wang, S., Lee, J. E., & Levy, B. R. (2020). Global reach of ageism on older persons' health: A systematic review. *PLOS ONE, 15* https://journals.plos.org/plosone/article?id =10.1371/journal.pone.0220857; Horton, S., Baker, J., & Deakin, J. M. (2007). Stereotypes of aging: Their effects on the health of seniors in North American society. *Educational Gerontology, 33,* 1021–1035; Meisner, B. A. (2012). A meta-analysis of positive and negative age stereotype priming effects on behavior among older adults. *The Journals of Gerontology, Series B: Psychological Sciences and Social Sciences, 67,* 13–17; Lamont, R. A., Swift, H. J., & Abrams, D. (2015). A review and meta-analysis of age-based stereotype threat: Negative stereotypes, not facts, do the damage. *Psychology and Aging, 30,* 180–193; Westerhof, G. J., Miche, M., Brothers, A. F., Barrett, A. E., Diehl, M., Montepare, J. M., . . . Wurm, S. (2014). The influence of subjective aging on health and longevity: A meta-analysis of longitudinal data. *Psychology and Aging, 29,* 793–802. https://doi.org/10.1037/a0038016.

11. Levy, B. R. (2009). Stereotype embodiment: A psychosocial approach to aging. *Current Directions in Psychological Science, 18,* 332–336; Kwong See, S. T., Rasmussen, C., & Pertman, S. Q. (2012). Measuring children's age stereotyping using a modified Piagetian conservation task. *Educational Gerontology, 38,* 149–165; Flamion, A., Missotten, P., Jennotte, L., Hody, N., & Adam, S. (2020). Old age-related stereotypes of preschool children. *Frontiers in Psychology, 11,* 807. https://doi.org/10.3389 /fpsyg.2020.00807.

12. Kwong See, S. T., Rasmussen, C., & Pertman, S. Q. (2012). Measuring children's age stereotyping using a modified Piagetian conservation task. *Educational Gerontology, 38,* 149–165; Flamion, A., Missotten, P., Jennotte, L., Hody, N., & Adam, S. (2020). Old age-related stereo-

types of preschool children. *Frontiers in Psychology, 11,* 807. https://doi
.org/10.3389/fpsyg.2020.00807.

13. Montepare, J. M., & Zebrowitz, L. A. (2002). A social-developmental
view of ageism. In T. D. Nelson (Ed.), *Ageism: Stereotyping and prejudice
against older persons* (pp. 77–125). Cambridge, MA: The MIT Press.

14. Officer, A., & de la Fuente-Núñez, V. (2018). A global campaign to combat ageism. *Bulletin of the World Health Organization, 96,* 295–296.

15. Bigler, R. S., & Liben, L. S. (2007). Developmental intergroup theory:
Explaining and reducing children's social stereotyping and prejudice.
Current Directions in Psychological Science, 16, 162–166.

16. Levy, B. R., Pilver, C., Chung, P. H., & Slade, M. D. (2014). Subliminal
strengthening: Improving older individuals' physical function over time
with an implicit-age-stereotype intervention. *Psychological Science, 25,*
2127–2135; Hausdorff, J. M., Levy, B., & Wei, J. (1999). The power of
ageism on physical function of older persons: Reversibility of age-related
gait changes. *Journal of the American Geriatrics Society, 47,* 1346–1349;
Levy, B. (2000). Handwriting as a reflection of aging self-stereotypes.
Journal of Geriatric Psychiatry, 33, 81–94.

17. Levy, B. (1996). Improving memory in old age through implicit self-
stereotyping. *Journal of Personality and Social Psychology, 71,* 1092–1107;
Levy, B. R., Zonderman, A. B., Slade, M. D., & Ferrucci, L. (2012).
Memory shaped by age stereotypes over time. *The Journals of Gerontology, Series B: Psychological Sciences and Social Sciences, 67,* 432–436.

18. Levy, B. R. (2009). Stereotype embodiment: A psychosocial approach to
aging. *Current Directions in Psychological Science, 18,* 332–336.

19. Levy, B. R., & Myers, L. M. (2004). Preventive health behaviors influenced by self-perceptions of aging. *Preventive Medicine, 39,* 625–629;
Levy, B. R., & Slade, M. D. (2019). Positive views of aging reduce risk
of developing later-life obesity. *Preventive Medicine Report, 13,* 196–198.

20. Levy, B. R., & Bavishi, A. (2018). Survival-advantage mechanism: Inflammation as a mediator of positive self-perceptions of aging on longevity. *The Journals of Gerontology, Series B: Psychological Sciences and Social
Sciences, 73,* 409–412; Levy, B. R., Moffat, S., Resnick, S. M., Slade,
M. D., & Ferrucci, L. (2016). Buffer against cumulative stress: Positive
age self-stereotypes predict lower cortisol across 30 years. *GeroPsych: The
Journal of Gerontopsychology and Geriatric Psychiatry, 29,* 141–146.

21. Epel, E. S., Crosswell, A. D., Mayer, S. E., Prather, A. A., Slavich, G. M.,
Puterman, E., & Mendes, W. B. (2018). More than a feeling: A unified
view of stress measurement for population science. *Frontiers in Neuroen-*

docrinology, 49, 146–169; McEwen, B. S. (2013). The brain on stress: Toward an integrative approach to brain, body, and behavior. *Perspectives on Psychological Science, 8,* 673–675.

22. Steele, C. (2014). Stereotype threat and African-American student achievement. In D. B. Grusky (Ed.), *Social stratification: Class, race and gender in sociological perspective.* New York: Taylor & Francis; Steele, C. M., & Aronson, J. (1995). Stereotype threat and the intellectual test performance of African Americans. *Journal of Personality and Social Psychology, 69,* 797–811.

23. Davies, P. G., Spencer, S. J., & Steele, C. M. (2005). Clearing the air: Identity safety moderates the effects of stereotype threat on women's leadership aspirations. *Journal of Personality and Social Psychology, 88,* 276–287.

24. Chopik, W. J., & Giasson, H. L. (2017). Age differences in explicit and implicit age attitudes across the life span. *The Gerontologist, 57,* 169–177; Montepare, J. M., & Lachman, M. E. (1989). "You're only as old as you feel": Self-perceptions of age, fears of aging, and life satisfaction from adolescence to old age. *Psychology and Aging, 4,* 73–78.

25. Levy, B. R., Zonderman, A. B., Slade, M. D., & Ferrucci, L. (2011). Memory shaped by age stereotypes over time. *The Journals of Gerontology, Series B: Psychological Sciences and Social Sciences, 67,* 432–436; Levy, B. R., Slade, M. D., & Kasl, S. (2002). Longitudinal benefit of positive self-perceptions of aging on functioning health. *The Journals of Gerontology, Series B: Psychological Sciences and Social Sciences, 57,* 409–417.

26. Desta, Y. (2020, June 30). Carl Reiner and Mel Brooks had comedy's most iconic friendship. *Vanity Fair.* https://www.vanityfair.com/holly wood/2020/06/carl-reiner-mel-brooks-friendship.

27. Nimrod, G., & Berdychevsky, L. (2018). Laughing off the stereotypes: Age and aging in seniors' online sex-related humor. *The Gerontologist, 58,* 960–969.

28. Levy, B. R. (2009). Stereotype embodiment: A psychosocial approach to aging. *Current Directions in Psychological Science, 18,* 332–336; Levy, B. R., Pilver, C., Chung, P. H., & Slade, M. D. (2014). Subliminal strengthening: Improving older individuals' physical function over time with an implicit-age-stereotype intervention. *Psychological Science, 25,* 2127–2135; Ng, R., Allore, H. G., Trentalange, M., Monin, J. K., & Levy, B. R. (2015). Increasing negativity of age stereotypes across 200 years: Evidence from a database of 400 million words. *PLOS ONE, 10*(2), e0117086.

29. Bodner, E., Palgi, Y., & Wyman, M. (2018). Ageism in mental health assessment and treatment of older adults. In L. Ayalon & C. Tesch-Römer (Eds.), *Contemporary perspectives on ageism*. New York: Springer; Laidlaw, K., & Pachana, N. A. (2009). Ageing, mental health, and demographic change: Challenges for psychotherapists. *Professional Psychology: Research and Practice, 40,* 601–608.

30. Graham, J. (2019, May 30). A doctor speaks out about ageism in medicine. Kaiser Health News. https://khn.org/news/navigating-aging-a-doctor-speaks-out-about-ageism-in-medicine/.

31. Newport, F. (2015, January 26). Only a third of the oldest baby boomers in US still working. Gallup. https://news.gallup.com/poll/181292/third-oldest-baby-boomers-working.aspx.

32. Pelisson, A., & Hartmans, A. (2017). The average age of employees at all the top tech companies, in one chart. *Insider.* https://www.businessinsider.com/median-tech-employee-age-chart-2017–8.

33. Applewhite, A. (2016, September 3). You're how old? We'll be in touch. *The New York Times.* https://www.nytimes.com/2016/09/04/opinion/sunday/youre-how-old-well-be-in-touch.html.

34. Passarino, G., De Rango, F., & Montesanto, A. (2016). Human longevity: Genetics or lifestyle? It takes two to tango. *Immunity and Ageing, 13,* 12. https://doi.org/10.1186/s12979-016-0066-z; Vaupel, J. W., Carey, J. R., Christensen, K., Johnson, T. E., Yashin, A. I., Holm, N. V., . . . & Curtsinger, J. W. (1998). Biodemographic trajectories of longevity. *Science, 280,* 855–860.

Chapter 2: Anatomy of a Senior Moment

1. Safire, W. (1998, May 10). On language; Great moment in moments. *New York Times Magazine.* https://www.nytimes.com/1998/05/10/magazine/on-language-great-moments-in-moments.html.

2. Maxwell, K. (2021). Senior moment. *Macmillan Dictionary.* https://www.macmillandictionary.com/us/buzzword/entries/senior-moment.html.

3. James, W. (1892). The stream of consciousness. In *Psychology* (Chapter 11, p. 251). New York: World Publishing Company; Cherry, K. (2020). William James psychologist biography: The father of American psychology. Verywell Mind. Retrieved June 14, 2021, from https://www.verywellmind.com/william-james-biography-1842-1910-2795545.

4. Ballesteros, S., Kraft, E., Santana, S., & Tziraki, C. (2015). Maintaining older brain functionality: A targeted review. *Neuroscience & Biobehavioral Reviews, 55,* 453–477.

5. American Psychological Association. (2021). Memory and aging. https://www.apa.org/pi/aging/memory-and-aging.pdf; Arkowitz, H., & Lilienfeld, S. O. (2012, November 1). Memory in old age can be bolstered: Researchers have found ways to lessen age-related forgetfulness. *Scientific American.* https://www.scientificamerican.com/article/memory-in-old-age-can-be-bolstered/; Belleville, S., Gilbert, B., Fontaine, F., Gagnon, L., Ménard, É., & Gauthier, S. (2006). Improvement of episodic memory in persons with mild cognitive impairment and healthy older adults: Evidence from a cognitive intervention program. *Dementia and Geriatric Cognitive Disorders, 22,* 486–499; Haj, M. E., Fasotti, L., & Allain, P. (2015). Destination memory for emotional information in older adults. *Experimental Aging Research, 41,* 204–219; Nyberg, L., Maitland, S. B., Rönnlund, M., Bäckman, L., Dixon, R. A., Wahlin, Å., & Nilsson, L.-G. (2003). Selective adult age differences in an age-invariant multifactor model of declarative memory. *Psychology and Aging, 18,* 149–160; Pennebaker, J. W., & Stone, L. D. (2003). Words of wisdom: Language use over the life span. *Journal of Personality and Social Psychology, 85,* 291–301.

6. Levy, B. (1996). Improving memory in old age through implicit self-stereotyping. *Journal of Personality and Social Psychology, 71,* 1092–1107; Schaie, K. W., & Willis, S. L. (2010). The Seattle Longitudinal Study of Adult Cognitive Development. *International Society for the Study of Behavioural Development Bulletin, 57,* 24–29.

7. Levy, B. (1996). Improving memory in old age through implicit self-stereotyping. *Journal of Personality and Social Psychology, 71,* 1092–1107.

8. Levy, B., & Langer, E. (1994). Aging free from negative stereotypes: Successful memory in China and among the American Deaf. *Journal of Personality and Social Psychology, 66,* 989–997.

9. I use the convention described by Padden and Humphries of using the lowercase *d* or *deaf* to refer to the audiological condition in contrast to the uppercase *D* or *Deaf* to describe the linguistic and cultural community. Padden, C., & Humphries, T. (1990). *Deaf in America: Voices from a culture.* Cambridge, MA: Harvard University Press.

10. Lei, X., Strauss, J., Tian, M., & Zhao, Y. (2015). Living arrangements of the elderly in China: Evidence from the CHARLS national baseline. *China Economic Journal, 8,* 191–214; Levy, B., & Langer, E. (1994). Aging free from negative stereotypes: Successful memory in China and among the American Deaf. *Journal of Personality and Social Psychology, 66,* 989–997; Nguyen, A. L., & Seal, D. W. (2014). Cross-cultural com-

parison of successful aging definitions between Chinese and Hmong elders in the United States. *Journal of Cross-Cultural Gerontology, 29,* 153–171.

11. Becker, G. (1980). *Growing old in silence.* Berkeley: University of California Press.

12. Ibid.

13. Ibid.

14. Levy, B., & Langer, E. (1994). Aging free from negative stereotypes: Successful memory in China and among the American Deaf. *Journal of Personality and Social Psychology, 66,* 989–997.

15. Palmore, E. B. (1988). *Facts on aging quiz: A handbook of uses and results.* New York: Springer.

16. Levy, B. R. (2009). Stereotype embodiment: A psychosocial approach to aging. *Current Directions in Psychological Science, 18,* 332–336; Levy, B., & Langer, E. (1994). Aging free from negative stereotypes: Successful memory in China and among the American Deaf. *Journal of Personality and Social Psychology, 66,* 989–997.

17. Parshley, L. (2018, May 29). This man memorized a 60,000-word poem using deep encoding. *Nautilus.* Retrieved June 14, 2021, from https://nautil.us/blog/-this-man-memorized-a-60000_word-poem-using-deep-encoding.

18. Levy, B., & Langer, E. (1994). Aging free from negative stereotypes: Successful memory in China and among the American Deaf. *Journal of Personality and Social Psychology, 66,* 989–997; Padden, C., & Humphries, T. (1988). *Deaf in America: Voices from a culture.* Cambridge, MA: Harvard University Press.

19. Devine, P. G. (1989). Stereotypes and prejudice: Their automatic and controlled components. *Journal of Personality and Social Psychology, 56,* 5–18. Also see: Bargh, J. A., & Pietromonaco, P. (1982). Automatic information processing and social perception: The influence of trait information presented outside of conscious awareness on impression formation. *Journal of Personality and Social Psychology, 43,* 437–449.

20. Levy, B. (1996). Improving memory in old age through implicit self-stereotyping. *Journal of Personality and Social Psychology, 71,* 1092–1107. We did not determine the duration of the effects because we were not expecting such dramatic results; however, our long-term studies have shown that internalized age stereotypes can have an effect across years.

21. Lee, K. E., & Lee, H. (2018). Priming effects of age stereotypes on

memory of older adults in Korea. *Asian Journal of Social Psychology, 22,* 39–46.

22. Horton, S., Baker, J., & Deakin, J. M. (2007). Stereotypes of aging: Their effects on the health of seniors in North American society. *Educational Gerontology, 33,* 1021–1035; Meisner, B. A. (2012). A meta-analysis of positive and negative age stereotype priming effects on behavior among older adults. *The Journals of Gerontology, Series B: Psychological Sciences and Social Sciences, 67,* 13–17; Lamont, R. A., Swift, H. J., & Abrams, D. (2015). A review and meta-analysis of age-based stereotype threat: Negative stereotypes, not facts, do the damage. *Psychology and Aging, 30,* 180–193.

23. Levy, B. R., Zonderman, A. B., Slade, M. D., & Ferrucci, L. (2012). Memory shaped by age stereotypes over time. *The Journals of Gerontology, Series B: Psychological Sciences and Social Sciences, 67,* 432–436.

24. Levitin, D. J. (2020). *Successful aging: A neuroscientist explores the power and potential of our lives.* New York: Penguin Random House.

25. Cabeza, R., Anderson, N. D., Locantore, J. K., & McIntosh, A. R. (2002). Aging gracefully: Compensatory brain activity in high-performing older adults. *Neuroimage, 17,* 1394–1402; Gutchess, A. (2014). Plasticity of the aging brain: New directions in cognitive neuroscience. *Science, 346,* 579–582.

26. Arora, D. (1991). *All that the rain promises and more: A hip pocket guide to Western mushrooms.* Berkeley, CA: Ten Speed Press.

27. Mead, M. (1975). *Culture and commitment: A study of the generation gap.* Garden City, NY: Natural History Press.

28. Ibid.

29. Klein, C. (2016, September 23). DNA study finds Aboriginal Australians world's oldest civilization. History. https://history.com/news/dna-study-finds-aboriginal-australians-worlds-oldest-civilization.

30. Archibald-Binge, E., & Geraghty, K. (2020, April 24). "We treat them like gold": Aboriginal community rallies around elders. Tharawal Aboriginal Corporation. Retrieved June 14, 2021, from https://tacams.com.au/2020/04/24/we-treat-them-like-gold-aboriginal-community-rallies-around-elders/; Malcolm, L., & Willis, O. (2016, July 8). Songlines: The Indigenous memory code. *All in the mind.* ABC Radio. Retrieved June 14, 2021, from https://www.abc.net.au/radionational/programs/allinthemind/songlines-indigenous-memory-code/7581788; Curran, G., Barwick, L., Turpin, M., Walsh, F., & Laughren, M. (2019). Central Australian Aboriginal songs and biocultural knowledge: Evidence from

women's ceremonies relating to edible seeds. *Journal of Ethnobiology, 39,* 354–370; Korff, J. (2020, August 13). Respect for elders and culture. Creative Spirits. https://www.creativespirits.info/aboriginalculture/peo ple/respect-for-elders-and-culture.

31. Mead, M. (1970). *Culture and commitment: A study of the generation gap.* Garden City, NY: Natural History Press.

Chapter 3: Old and Fast

1. Levy, B. R., Slade, M. D., & Kasl, S. (2002). Longitudinal benefit of positive self-perceptions of aging on functioning health. *The Journals of Gerontology, Series B: Psychological Sciences and Social Sciences, 57,* 409–417.

2. Sargent-Cox, K., Anstey, K. J., & Luszcz, M. A. (2012). The relationship between change in self-perceptions of aging and physical functioning in older adults. *Psychology and Aging, 27,* 750–760; Ayalon, L. (2016). Satisfaction with aging results in reduced risk for falling. *International Psychogeriatrics, 28,* 741–747.

3. Hausdorff, J. M., Levy, B., & Wei, J. (1999). The power of ageism on physical function of older persons: Reversibility of age-related gait changes. *Journal of the American Geriatrics Society, 47,* 1346–1349.

4. Levy, B. R., Pilver, C., Chung, P. H., & Slade, M. D. (2014). Subliminal strengthening: Improving older individuals' physical function over time with an implicit-age-stereotype intervention. *Psychological Science, 25,* 2127–2135.

5. McAuley, E. W., Wójcicki, T. R., Gothe, N. P., Mailey, E. L., Szabo, A. N., Fanning, J., . . . & Mullen, S. P. (2013). Effects of a DVD-delivered exercise intervention on physical function in older adults. *The Journals of Gerontology, Series A: Biological Sciences and Medical Sciences, 68,* 1076–1082.

6. Levy, B. R., Pilver, C., Chung, P. H., & Slade, M. D. (2014). Subliminal strengthening: Improving older individuals' physical function over time with an implicit-age-stereotype intervention. *Psychological Science, 25,* 2127–2135. In social-psychological experiments, this snowballing has been found with other populations and is called a "recursive effect." It occurs when an intervention leads to a change in mindset, which leads to both an internal change and a change in one's interactions with the environment. These then reinforce the change in mindset. To read more about these concepts see: Walton, G. M., & Wilson, T. D. (2018). Wise interventions: Psychological remedies for social and personal problems. *Psychological Review, 125,* 617–655.

7. Levy, B. R., Slade, M., Chang, E. S., Kannoth, S., & Wang, S. H. (2020). Ageism amplifies cost and prevalence of health conditions. *The Gerontologist, 60,* 174–181; Ng, R., Allore, H. G., Trentalange, M., Monin, J. K., & Levy, B. R. (2015). Increasing negativity of age stereotypes across 200 years: Evidence from a database of 400 million words. *PLOS ONE, 10*(2), e0117086.

8. Rutemiller, B. (2018, July 30). 97-year-old Maurine Kornfeld to be inducted into international masters swimming hall of fame. *Swimming World.* https://www.swimmingworldmagazine.com/news/97-year-old -maurine-kornfeld-to-be-inducted-into-international-masters-swimming -hall-of-fame/.

9. Reynolds, G. (2019, September 18). Taking up running after 50? It's never too late to shine. *The New York Times.* https://www.nytimes. com/2019/09/18/well/move/taking-up-running-after-50-its-never -too-late-to-shine.html; Piasecki, J., Ireland, A., Piasecki, M., Deere, K., Hannam, K., Tobias, J., & McPhee, J. S. (2019). Comparison of muscle function, bone mineral density and body composition of early starting and later starting older Masters athletes. *Frontiers in Physiology, 10,* 1050.

10. Hardy, S. E., & Gill, T. M. (2004). Recovery from disability among community-dwelling older persons. *JAMA, 291,* 1596–1602.

11. Levy, B. R., Slade, M. D., Murphy, T. E., & Gill, T. M. (2012). Association between positive age stereotypes and recovery from disability in older persons. *JAMA, 308,* 1972–1973.

12. Glaister, D. (2008, August 4). Hollywood actor Morgan Freeman seriously hurt in crash. *The Guardian.* https://www.theguardian.com /film/2008/aug/04/morgan.freeman.seriously.hurt.in.crash.

13. Pringle, G. (2020, January 22). Q&A: Just getting started with Morgan Freeman. Senior Planet. https://seniorplanet.org/just-getting-started/.

14. Durocher, K. (2016, February 21). Morgan Freeman opens up about aging in Hollywood and what he thinks it takes to be president. [Video]. YouTube. https://www.youtube.com/watch?v=5viYDBzpuoQ.

Chapter 4: Brawny Brains: Genes Aren't Destiny

1. Maurer, K., Volk, S., & Gerbaldo, H. (1997). Auguste D and Alzheimer's disease. *Lancet, 349,* 1546–1549.

2. Ibid.

3. Samanez-Larkin, G. R. (2019). *The aging brain: Functional adaptation across the lifespan.* Washington, DC: American Psychological Associa-

tion; Butler, R. N. (2008). *The longevity revolution: The benefits and challenges of living a long life*. New York: PublicAffairs.

4. Grant, W. B., Campbell, A., Itzhaki, R. F., & Savory, J. (2002). The significance of environmental factors in the etiology of Alzheimer's disease. *Journal of Alzheimer's Disease, 4,* 179–189.

5. In a study of twenty-six countries, it was found that mainland China and India had the most-positive societal views toward aging. Löckenhoff, C. E., De Fruyt, F., Terracciano, A., McCrae, R., De Bolle, M., . . . & Yik, M. (2009). Perceptions of aging across 26 cultures and their culture-level associates. *Psychology and Aging, 24,* 941–954; Chahda, N. K. (2012). Intergenerational relationships: An Indian perspective. University of Delhi. United Nations Department of Economic and Social Affairs Family. https://www.un.org/esa/socdev/family/docs/egm12/CHADHA-PAPER.pdf/.

6. Cohen, L. (2000). *No aging in India*. Berkeley: University of California Press, p. 17.

7. Levy, B. R., Ferrucci, L., Zonderman, A. B., Slade, M. D., Troncoso, J., & Resnick, S. M. (2016). A culture–brain link: Negative age stereotypes predict Alzheimer's disease biomarkers. *Psychology and Aging, 31,* 82–88.

8. Bedrosian, T. A., Quayle, C., Novaresi, N., & Gage, F. H. (2018). Early life experience drives structural variation of neural genomes in mice. *Science, 359,* 1395–1399.

9. Weiler, N. (2017, January 9). Cultural differences may leave their mark on DNA. University of California San Francisco. https://www.ucsf.edu/news/2017/01/405466/cultural-differences-may-leave-their-mark-dna; Galanter, J. M., Gignoux, C. R., Oh, S. S., Torgerson, D., Pino-Yanes, M., Thakur, N., . . . & Zaitlen, N. (2017). Differential methylation between ethnic sub-groups reflects the effect of genetic ancestry and environmental exposures. *eLife, 6,* e20532. https://doi.org/10.7554/eLife.20532; Nishimura, K. K., Galanter, J. M., Roth, L. A., Oh, S. S., Thakur, N., Eng, C., . . . & Burchard, E. G. (2013). Early-life air pollution and asthma risk in minority children. The GALA II and SAGE II studies. *American Journal of Respiratory and Critical Care Medicine, 188,* 309–318.

10. Levy, B. R., Slade, M. D., Pietrzak, R. H., & Ferrucci, L. (2018). Positive age beliefs protect against dementia even among elders with high-risk gene. *PLOS ONE, 13,* e191004. https://doi.org/10.1371/journal.pone.0191004.

11. Sperling, R. A., Donohue, M. C., Raman, R., Chung-Kai, S., Yaari, R., . . . A4 Study Team. (2020). Association of factors with ele-

vated amyloid burden in clinically normal older individuals. *JAMA Neurology, 77,* 735–745. Names and details of A4 participants interviewed have been changed to protect their privacy.

12. Justice, N. J. (2018). The relationship between stress and Alzheimer's disease. *Neurobiology of Stress, 8,* 127–133.

13. Brody, J. (2019, December 23). Tackling inflammation to fight age-related ailments. *The New York Times.* https://www.nytimes.com/2019/12/23/well/live/inflammation-aging-age-heart-disease-cancer-alzheimers-dementia-diabetes-depression-health.html; Epel, E. S., Crosswell, A. D., Mayer, S. E., Prather, A. A., Slavich, G. M., Puterman, E., & Mendes, W. B. (2018). More than a feeling: A unified view of stress measurement for population science. *Frontiers in Neuroendocrinology, 49,* 146–169; McEwen, B. S. (2013). The brain on stress: Toward an integrative approach to brain, body, and behavior. *Perspectives on Psychological Science, 8,* 673–675.

14. Levy, B. R., Hausdorff, J. M., Hencke, R., & Wei, J. Y. (2000). Reducing cardiovascular stress with positive self-stereotypes of aging. *The Journals of Gerontology, Series B: Psychological Sciences and Social Sciences, 55,* 205–213.

15. Attie, B., & Goldwater, J. (Directors). (2003). *Maggie growls.* Independent Lens. PBS. pbs.org/independentlens/maggiegrowls/index.

16. Levy, B. R., Slade, M. D., Pietrzak, R. H., & Ferrucci, L. (2020). When culture influences genes: Positive age beliefs amplify the cognitive-aging benefit of *APOE* ε2. *The Journals of Gerontology, Series B: Psychological Sciences and Social Sciences, 75,* e198–e203.

17. Suri, S., Heise, V., Trachtenberg, A. J., & Mackay, C. E. (2013). The forgotten *APOE* allele: A review of the evidence and suggested mechanisms for the protective effect of *APOE* ε2. *Neuroscience and Biobehavioral Reviews, 37,* 2878–2886.

18. Levy, B. R., Slade, M. D., Pietrzak, R. H., & Ferrucci, L. (2020). When culture influences genes: Positive age beliefs amplify the cognitive-aging benefit of *APOE* ε2. *The Journals of Gerontology, Series B: Psychological Sciences and Social Sciences, 75,* e198–e203.

19. Samanez-Larkin, G. R. (2019). *The aging brain: Functional adaptation across the lifespan.* Washington, DC: American Psychological Association.

20. Nottebohm, F. (2010). Discovering nerve cell replacement in the brains of adult birds. The Rockefeller University. http://centennial.rucares.org/index.php?page=Brain_Generates_Neurons.

21. Galvan, V., & Jin, K. (2007). Neurogenesis in the aging brain. *Clinical Interventions in Aging, 2*, 605–610.

22. Eriksson, P. S., Perfilieva, E., Björk-Eriksson, T., Alborn, A. M., Nordborg, C., Peterson, D. A., & Gage, F. H. (1998). Neurogenesis in the adult human hippocampus. *Nature Medicine, 4*, 1313–1317; Porto, F. H., Fox, A. M., Tusch, E. S., Sorond, F., Mohammed, A. H., & Daffner, K. R. (2015). In vivo evidence for neuroplasticity in older adults. *Brain Research Bulletin, 114*, 56–61.

23. Müller, P., Rehfeld, K., Schmicker, M., Hökelmann, A., Dordevic, M., Lessmann, V., Brigadski, T., Kaufmann, J., & Müller, N. G. (2017). Evolution of neuroplasticity in response to physical activity in old age: The case for dancing. *Frontiers in Aging Neuroscience, 9*, 56. https://doi.org/10.3389/fnagi.2017.00056; Park, D. C., & Bischof, G. N. (2011). Neuroplasticity, aging, and cognitive function. In K. W. Schaie, S. Willis, B. G. Knight, B. Levy, & D. C. Park (Eds.), *Handbook of the psychology of aging* (7th ed., pp. 109–119). New York: Elsevier.

Chapter 5: Later-Life Mental Health Growth

1. Levy, B. R., Hausdorff, J. M., Hencke, R., & Wei, J. Y. (2000). Reducing cardiovascular stress with positive self-stereotypes of aging. *The Journals of Gerontology, Series B: Psychological Sciences and Social Sciences, 55*, 205–213.

2. Levy, B. R., Moffat, S., Resnick, S. M., Slade, M. D., & Ferrucci, L. (2016). Buffer against cumulative stress: Positive age self-stereotypes predict lower cortisol across 30 years. *GeroPsych: The Journal of Gerontopsychology and Geriatric Psychiatry, 29*, 141–146.

3. Ibid.

4. Levy, B. R., Chung, P. H., Slade, M. D., Van Ness, P. H., & Pietrzak, R. H. (2019). Active coping shields against negative aging self-stereotypes contributing to psychiatric conditions. *Social Science and Medicine, 228*, 25–29.

5. Levy, B. R., Pilver, C. E., & Pietrzak, R. H. (2014). Lower prevalence of psychiatric conditions when negative age stereotypes are resisted. *Social Science and Medicine, 119*, 170–174.

6. American Psychological Association. (2014). Guidelines for psychological practice with older adults. *American Psychologist, 69*, 34–65; Hinrichsen, G. A. (2015). Attitudes about aging. In P. A. Lichtenberg, B. T. Mast, B. D. Carpenter, & J. Loebach Wetherell (Eds.), *APA handbook of clinical geropsychology* (pp. 363–377). Washington, DC: American Psy-

chological Association; Robb, C., Chen, H., & Haley, W. E. (2002). Ageism in mental health and health care: A critical review. *Journal of Clinical Geropsychology, 8,* 1–12.

7. World Health Organization. (2017, December 12). Mental health of older adults. https://www.who.int/news-room/fact-sheets/detail/men tal-health-of-older-adults; Segal, D. L., Qualls, S. H., & Smyer, M. A. (2018). *Aging and mental health.* Hoboken, NJ: Wiley Blackwell.

8. Freud, S. (1976). On psychotherapy. In *The complete psychological works of Sigmund Freud.* New York: W. W. Norton, p. 264.

9. Woodward, K. M. (1991). *Aging and its discontents: Freud and other fictions.* Bloomington: Indiana University Press, p. 3.

10. Abend, S. M. (2016). *A brief introduction to Sigmund Freud's psychoanalysis and his enduring legacy.* Astoria, NY: International Psychoanalytic Books.

11. Ibid.

12. Helmes, E., & Gee, S. (2003). Attitudes of Australian therapists toward older clients: Educational and training imperatives. *Educational Gerontology, 29,* 657–670.

13. Hinrichsen, G. A. (2015). Attitudes about aging. In P. A. Lichtenberg, B. T. Mast, B. D. Carpenter, & J. Loebach Wetherell (Eds.), *APA handbook of clinical geropsychology* (pp. 363–377). Washington, DC: American Psychological Association; Helmes, E., & Gee, S. (2003). Attitudes of Australian therapists toward older clients: Educational and training imperatives. *Educational Gerontology, 29,* 657–670.

14. Cuijpers, P., Karyotaki, E., Eckshtain, D., Ng, M. Y., Corteselli, K. A., Noma, H., Quero, S., & Weisz, J. R. (2020). Psychotherapy for depression across different age groups: A systematic review and meta-analysis. *JAMA Psychiatry, 77,* 694–702.

15. Plotkin, D. (2014). Older adults and psychoanalytic treatment: It's about time. *Psychodynamic Psychiatry, 42,* 23–50; Chen, Y., Peng, Y., & Fang, P. (2016). Emotional intelligence mediates the relationship between age and subjective well-being. *International Journal of Aging & Human Development, 83,* 91–107; Funkhouser, A. T., Hirsbrunner, H., Cornu, C., & Bahro, M. (1999). Dreams and dreaming among the elderly: An overview. *Aging & Mental Health, 3*(1), 10–20; O'Rourke, N., Cappeliez, P., & Claxton, A. (2011). Functions of reminiscence and the psychological well-being of young and older adults over time. *Aging & Mental Health, 15,* 272–281.

16. American Psychological Association. (2020). APA resolution on ageism.

Washington, DC: American Psychological Association. https://www
.apa.org/about/policy/resolution-ageism.pdf.

17. Moye, J., Karel, M. J., Stamm, K. E., Qualls, S. H., Segal, D. L., Tazeau,
Y. N., & DiGilio, D. A. (2019). Workforce analysis of psychological
practice with older adults: Growing crisis requires urgent action. *Training and Education in Professional Psychology, 13,* 46–55.

18. Cuijpers, P., Sijbrandij, M., Koole, S. L., Andersson, G., Beekman,
A. T., & Reynolds, C. F. (2014). Adding psychotherapy to antidepressant medication in depression and anxiety disorders: A meta-analysis.
World Psychiatry: Official Journal of the World Psychiatric Association, 13,
56–67; Reynolds, C. F., Frank, E., Perel, J. M., Imber, S. D., Cornes,
C., Miller, M. D., . . . & Kupfer, D. J. (1999). Nortriptyline and interpersonal psychotherapy as maintenance therapies for recurrent major depression: A randomized controlled trial in patients older than 59 years.
JAMA, 281, 39–45; Sammons, M. T., & McGuinness, K. M. (2015).
Combining psychotropic medications and psychotherapy generally leads
to improved outcomes and therefore reduces the overall cost of care. Society for Prescribing Psychology. https://www.apadivisions.org/division
-55/publications/tablet/2015/04/combininations.

19. Grand View Research. (2020, March). U.S. long term care market size,
share and trends analysis by service, and segment forecasts. https://
www.grandviewresearch.com/industry-analysis/us-long-term-care-ltc
-market.

20. Brown, B. (2018, April 24). Bethany Brown discusses human rights
violations in US nursing homes. Yale Law School. https://law.yale.edu
/yls-today/news/bethany-brown-discusses-human-rights-violations-us
-nursing-homes; Human Rights Watch. (2018). "They want docile":
How nursing homes in the United States overmedicate people with dementia. https://www.hrw.org/report/2018/02/05/they-want-docile/how
-nursing-homes-united-states-overmedicate-people-dementia#_ftn64);
Ray, W. A., Federspiel, C. F., & Schaffner, W. (1980). A study of antipsychotic drug use in nursing homes: Epidemiologic evidence suggesting
misuse. *American Journal of Public Health, 70,* 485–491.

21. Human Rights Watch. (2018). "They want docile": How nursing homes
in the United States overmedicate people with dementia. https://www
.hrw.org/report/2018/02/05/they-want-docile/how-nursing-homes-unit
ed-states-overmedicate-people-dementia#_ftn64).

22. Hinrichsen, G. A. (2015). Attitudes about aging. In P. A. Lichtenberg,
B. T. Mast, B. D. Carpenter, & J. Loebach Wetherell (Eds.), *APA hand-*

book of clinical geropsychology (pp. 363–377). Washington, DC: American Psychological Association; Park, M., & Unützer, J. (2011). Geriatric depression in primary care. *The Psychiatric Clinics of North America, 34,* 469–487.

23. Axelrod, J., Balaban, S., & Simon, S. (2019, July 27). Isolated and struggling, many seniors are turning to suicide. NPR. https://www .npr.org/2019/07/27/745017374/isolated-and-struggling-many-seniors -are-turning-to-suicide; Conwell, Y., Van Orden, K., & Caine, E. D. (2011). Suicide in older adults. *The Psychiatric Clinics of North America, 34,* 451–468. https://doi.org/10.1016/j.psc.2011.02.002; Canetto, S. S. (2017). Suicide: Why are older men so vulnerable? *Men and Masculinities, 20,* 49–70.

24. Park, M., & Unützer, J. (2011). Geriatric depression in primary care. *The Psychiatric Clinics of North America, 34,* 469–487; Span, P. (2020, October 30). You're not too old to talk to someone. *The New York Times.* https://www.nytimes.com/2020/10/30/health/mental-health-psycho therapy-elderly.html.

25. Span, P. (2020, October 30). You're not too old to talk to someone. *The New York Times.* https://www.nytimes.com/2020/10/30/health /mental-health-psychotherapy-elderly.html.

26. Coles, R. (1970, October 31). I-The measure of man. *The New Yorker.* https://www.newyorker.com/magazine/1970/11/07/i-the-measure-of -man.

27. Erikson, E. (1993). *Gandhi's truth: On the origins of militant nonviolence.* New York: W. W. Norton.

28. Erikson, E. H., Erikson, J. M., & Kivnick, H. Q. (1989). *Vital involvement in old age.* New York: W. W. Norton.

29. Goleman, D. (1988, June 14). Erikson, in his own old age, expands his view of life. *The New York Times.* https://www.nytimes.com/1988/06/14 /science/erikson-in-his-own-old-age-expands-his-view-of-life.html.

30. Ibid.

31. Ibid.

32. Cowan, R., & Thal, L. (2015). *Wise aging: Living with joy, resilience, & spirit.* Springfield, NJ: Behrman House.

33. Chibanda, D., Weiss, H. A., Verhey, R., Simms, V., Munjoma, R., Rusakaniko, S., . . . & Araya, R. (2016). Effect of a primary care–based psychological intervention on symptoms of common mental disorders in Zimbabwe: A randomized clinical trial. *JAMA, 316,* 2618–2626.

Chapter 6: Longevity Advantage of 7.5 Years

1. Atchley, R. C. (1999). *Continuity and adaptation in aging: Creating positive experiences.* Baltimore: Johns Hopkins University Press.
2. Levy, B. R., Slade, M. D., Kunkel, S. R., & Kasl, S. V. (2002). Longevity increased by positive self-perceptions of aging. *Journal of Personality and Social Psychology, 83,* 261–270. This survival-advantage figure is based on the difference in the amount of time it took half of the people to die with either positive or negative age beliefs.
3. Ibid., p. 268.
4. The image of aging in media and marketing. (2002, September 4). Hearing before the US Senate Special Committee on Aging. https://www.govinfo.gov/content/pkg/CHRG-107shrg83476/html/CHRG-107shrg83476.htm.
5. Ibid., Doris Roberts testimony.
6. Officer, A., & de la Fuente-Núñez, V. (2018). A global campaign to combat ageism. *Bulletin of the World Health Organization, 96,* 295–296; Kotter-Grühn, D., Kleinspehn-Ammerlahn, A., Gerstorf, D., & Smith, J. (2009). Self-perceptions of aging predict mortality and change with approaching death: 16-year longitudinal results from the Berlin Aging Study. *Psychology and Aging, 24,* 654–667; Sargent-Cox, K. A., Anstey, K. J., & Luszcz, M. A. (2014). Longitudinal change of self-perceptions of aging and mortality. *The Journals of Gerontology, Series B: Psychological Sciences and Social Sciences, 69,* 168–73; Zhang, X., Kamin, S. T., Liu, S., Fung, H. H., & Lang, F. R. (2018). Negative self-perception of aging and mortality in very old Chinese adults: The mediation role of healthy lifestyle. *The Journals of Gerontology, Series B: Psychological Sciences and Social Sciences, 75,* 1001–1009.
7. Passarino, G., De Rango, F., & Montesanto, A. (2016). Human longevity: Genetics or lifestyle? It takes two to tango. *Immunity and Ageing, 13,* 12. https://doi.org/10.1186/s12979-016-0066-z; Vaupel, J. W., Carey, J. R., Christensen, K., Johnson, T. E., Yashin, A. I., Holm, N. V., . . . & Curtsinger, J. W. (1998). Biodemographic trajectories of longevity. *Science, 280,* 855–860.
8. Levy, B. R., Slade, M. D., Pietrzak, R. H., & Ferrucci, L. (2020). When culture influences genes: Positive age beliefs amplify the cognitive-aging benefit of *APOE* ε2. *The Journals of Gerontology, Series B: Psychological Sciences and Social Sciences, 75,* e198–e203. https://doi.org/10.1093/geronb/gbaa126.

9. Hjelmborg, J., Iachine, I., Skytthe, A., Vaupel, J. W., McGue, M., Koskenvuo, M., Kaprio, J., Pedersen, N. L., & Christensen, K. (2006). Genetic influence on human lifespan and longevity. *Human Genetics, 119*, 312–321; Guillermo Martínez Corrales, G., & Nazif, A. (2020). Evolutionary conservation of transcription factors affecting longevity. *Trends in Genetics, 36*, 373–382.

10. Govindaraju, D., Atzmon, G., & Barzilai, N. (2015). Genetics, lifestyle and longevity: Lessons from centenarians. *Applied & Translational Genomics, 4*, 23–32.

11. Kearney, H. (2019). *QueenSpotting: Meet the remarkable queen bee and discover the drama at the heart of the hive.* North Adams, MA: Storey Publishing.

12. Levy, B. R. (2009). Stereotype embodiment: A psychosocial approach to aging. *Current Directions in Psychological Science, 18*, 332–336.

13. Idler, E. L., & Kasl, S. V. (1992). Religion, disability, depression, and the timing of death. *American Journal of Sociology, 97*, 1052–1079.

14. Levy, B., Ashman, O., & Dror, I. (2000). To be or not to be: The effects of aging stereotypes on the will to live. *Omega: Journal of Death and Dying, 40*, 409–420.

15. Levy, B. R., Slade, M. D., Kunkel, S. R., & Kasl, S. V. (2002). Longevity increased by positive self-perceptions of aging. *Journal of Personality and Social Psychology, 83*, 261–270.

16. Levy, B. R., & Bavishi, A. (2018). Survival-advantage mechanism: Inflammation as a mediator of positive self-perceptions of aging on longevity. *The Journals of Gerontology, Series B: Psychological Sciences and Social Sciences, 73*, 409–412.

17. Harris, T. B., Ferrucci, L., Tracy, R. P., Corti, M. C., Wacholder, S., Ettinger, W. H., & Wallace, R. (1999). Associations of elevated interleukin-6 and C-reactive protein levels with mortality in the elderly. *The American Journal of Medicine, 106*, 506–512; Szewieczek, J., Francuz, T., Dulawa, J., Legierska, K., Hornik, B., Włodarczyk, I., & Batko-Szwaczka, A. (2015). Functional measures, inflammatory markers and endothelin-1 as predictors of 360-day survival in centenarians. *Age, 37*, 85. https://doi.org/10.1007/s11357-015-9822-9.

18. Levy, B., Ashman, O., Dror, I. (2000). To be or not to be: The effects of aging stereotypes on the will to live. *Omega: Journal of Death and Dying, 40*, 409–420.

19. Levy, B. R., Provolo, N., Chang, E.-S., & Slade, M. D. (2021). Negative age stereotypes associated with older persons' rejection of

COVID-19 hospitalization. *Journal of the American Geriatrics Society,* *69,* 317–318.

20. Cole, T. R. (1993). *The journey of life: A cultural history of aging in America.* New York: Cambridge University Press.

21. Aronson, L. (2019). *Elderhood: Redefining aging, transforming medicine, reimagining life.* New York: Bloomsbury Publishing; Butler, R. N. (2011). *The longevity prescription: The 8 proven keys to a long, healthy life.* New York: Avery.

22. Butler, R. N. (2008). *The longevity revolution: The benefits and challenges of living a long life.* New York: PublicAffairs, p. xi.

23. Vaupel, J. W., Villavicencio, F., & Bergeron-Boucher, M. (2021). Demographic perspectives on the rise of longevity. *Proceedings of the National Academy of Sciences, 118 (9),* e2019536118. https://doi.org/10.1073/pnas .2019536118; Oeppen, J., & Vaupel, J. W. (2002). Broken limits to life expectancy. *Science, 296,* 1029–1031.

24. Şahin, D. B., & Heiland, F. W. (2016). Black-white mortality differentials at old-age: New evidence from the national longitudinal mortality study. *Applied demography and public health in the 21st century* (pp. 141–162). New York: Springer.

25. McCoy, R. (2011). African American elders, cultural traditions, and family reunions. *Generations: American Society on Aging.* https://gener ations.asaging.org/african-american-elders-traditions-family-reunions.

26. Mitchell, G. W. (2014). The silver tsunami. *Physician Executive, 40,* 34–38.

27. Wilkerson, I. (2020). *Caste: The origins of our discontents.* New York: Random House; Hacker, J. S., & Pierson, P. (2020). *Let them eat tweets: How the right rules in an age of extreme inequality.* New York: W. W. Norton.

28. Zarroli, J. (2020, December 9). Soaring stock market creates a club of centibillionaires. *All Things Considered.* https://www.npr.org /2020/12/09/944739703/soaring-stock-market-creates-a-club-of-centi billionaires.

29. Fried, L. (2021, May 6). Combating loneliness in aging: Toward a 21st century blueprint for societal connectedness. Age Boom Academy. The Robert N. Butler Columbia Aging Center. New York. Columbia University.

30. Elgar, F. J. (2010). Income inequality, trust, and population health in 33 countries. *American Journal of Public Health, 100,* 2311–2315.

31. Coughlin, J. (2017). *The longevity economy: Unlocking the world's fastest-growing, most misunderstood market.* New York: PublicAffairs.

32. Dychtwald, K. (2014). Longevity market emerges. In P. Irving (Ed.), *The upside of aging: How long life is changing the world of health, work, innovation, policy, and purpose*. New York: Wiley. It should also be noted that this same age group has higher poverty levels than any other. That is, there is a considerable wealth disparity among the older population.

33. Lee, R. (2020). Population aging and the historical development of inter-generational transfer systems. *Genus, 76*, 31. https://genus.springeropen.com/articles/10.1186/s41118-020-00100-8.

34. Azoulay, P., Jones, B. F., Kim, J. D., & Miranda, J. (2020). Age and high-growth entrepreneurship. *American Economic Review, 2*, 65–82. https://www.aeaweb.org/articles?id=10.1257/aeri.20180582.

35. Butler, R. N. (2008). *The longevity revolution: The benefits and challenges of living a long life*. New York: PublicAffairs; Roser, M., Ortiz-Ospina, E., & Ritchie, H. (2013). Life expectancy. Our World in Data. Retrieved June 1, 2021, from https://ourworldindata.org/life-expectancy.

36. Willigen, J., & Lewis, D. (2006). Culture as the context of aging. In H. Yoon and J. Hendricks (Eds.), *Handbook of Asian aging*. Amityville, NY: Baywood, p. 133.

37. Fries, J. F. (2000). Compression of morbidity in the elderly. *Vaccine, 18*, 1584–1589.

38. Butler, R. N. (2011). *The longevity prescription: The 8 proven keys to a long, healthy life*. New York: Avery.

39. Levy, B. R. (2017). Age-stereotype paradox: A need and opportunity for social change. *The Gerontologist, 57*, 118–126; Kolata, G. (2016, July 8). A medical mystery of the best kind: Major diseases are in decline. *The New York Times*. https://www.nytimes.com/2016/07/10/upshot/a-medical-mystery-of-the-best-kind-major-diseases-are-in-decline.html; Schoeni, R. F., Freedman, V. A., & Martin, L. M. (2008). Why is late-life disability declining? *Milbank Quarterly, 86*, 47–89.

40. Butler, R. N. (2011). *The longevity prescription: The 8 proven keys to a long, healthy life*. New York: Avery.

41. Andersen, S. L., Sebastiani, P., Dworkis, D. A., Feldman, L., & Perls, T. T. (2012). Health span approximates life span among many super-centenarians: Compression of morbidity at the approximate limit of life span. *The Journals of Gerontology, Series A: Biological Sciences and Medical Sciences, 67*, 395–405.

42. Mahoney, D., & Restak, R. (1999). *The longevity strategy*. New York: Wiley.

43. Schoenhofen, E. A., Wyszynski, D. F., Andersen, S., Pennington, J.,

Young, R., Terry, D. F., & Perls, T. T. (2006). Characteristics of 32 super-centenarians. *Journal of the American Geriatrics Society, 54,* 1237–1240.

44. Brody, J. (2021, June 21). The secrets of "cognitive super-agers." *The New York Times.* https://www.nytimes.com/2021/06/21/well/mind/aging-memory-centenarians.html; Beker, N., Ganz, A., Hulsman, M., Klausch, T., Schmand, B. A., Scheltens, P., Sikkes, S. A., & Holstege, H. (2021). Association of cognitive function trajectories in centenarians with postmortem neuropathology, physical health, and other risk factors for cognitive decline. *JAMA Network Open, 4*(1): e2031654. doi:10.1001/jamanetworkopen.2020.31654.

45. Nuwer, R. (2015, March 31). Lessons of the world's most unique super-centenarians. BBC. https://www.bbc.com/future/article/20150331-the-most-unique-supercentenarians.

46. Bucholz, K. (2021, February 5). Where 100 is the new 80. Statista. https://www.statista.com/chart/14931/where-100-is-the-new-80/; Santos-Lozano, A., Sanchis-Gomar, F., Pareja-Galeano, H., Fiuza-Luces, C., Emanuele, E., Lucia, A., & Garatachea, N. (2015). Where are super-centenarians located? A worldwide demographic study. *Rejuvenation Research, 18,* 14–19.

47. Sorezore nansai? Choju iwai no kisochishiki (How old? Basics of longevity celebrations). (2014, March 7). Gift Concierge. https://www.ringbell.co.jp/giftconcierge/2201.

48. Saikourei 116sai "imagatanoshii" Fukuoka no Tanaka san, Guiness nintei (116 years old, longest-living Ms. Tanaka of Fukuoka, recognized by Guinness World Records). (2019, March 10). *Nishinopponshinbun.* https://www.nishinippon.co.jp/item/n/492939/.

49. World's oldest person marks 118th birthday in Fukuoka. (2021, January 2). *The Japan Times.* https://www.japantimes.co.jp/news/2021/01/02/national/worlds-oldest-person-marks-118th-birthday-fukuoka/; Hanada, M. (2010). *Hana mo Arashi mo Hyaku Nanasai Tanaka Kane Choju Nihon Ichi heno Chosen* (107 year-old flower and storm, Kane Tanaka, the road to becoming the longest-living person in Japan). Fukuoka, Japan: Azusa Shoin.

50. Mizuno, Y. (1991). *Nihon no Bungaku to Oi* (Japan's literature and aging). Shitensha, 2.

51. Ackerman, L. S., & Chopik, W. J. (2021). Cross-cultural comparisons in implicit and explicit age bias. *Personality and Social Psychology Bulletin, 47,* 953–968.

52. Markus, H. R., & Kitayama, S. (1991). Culture and the self: Implica-

tions for cognition, emotion, and motivation. *Psychological Review, 98,* 224–253.

53. Ibid.

54. Ackerman, L. S., & Chopik, W. J. (2021). Cross-cultural comparisons in implicit and explicit age bias. *Personality and Social Psychology Bulletin, 47,* 953–968.

55. Levy, B. R., & Bavishi, A. (2018). Survival-advantage mechanism: Inflammation as a mediator of positive self-perceptions of aging on longevity. *The Journals of Gerontology, Series B: Psychological Sciences and Social Sciences, 73,* 409–412; Levy, B. R., Slade, M. D., Kunkel, S. R., & Kasl, S. V. (2002). Longevity increased by positive self-perceptions of aging. *Journal of Personality and Social Psychology, 83,* 261–270.

56. This is based on social psychologist Kurt Lewin's human behavior equation, $B = f(P, E)$. Lewin, K. (1936). *Principles of topological psychology.* New York: McGraw-Hill.

Chapter 7: Stars Invisible by Day: Creativity and the Senses

1. Doherty, M. J., Campbell, N. M., Tsuji, H., & Phillips, W. A. (2010). The Ebbinghaus illusion deceives adults but not young children. *Developmental Science, 13,* 714–721.

2. Schudel, M. (2016, June 7). Jerome S. Bruner, influential psychologist of perception, dies at 100. *Washington Post.* https://www.washingtonpost.com/national/jerome-s-bruner-influential-psychologist-of-perception-dies-at-100/2016/06/07/033e5870-2cc3-11e6-9b37-42985f6a265c_story.html.

3. Bruner, J. S., & Goodman, C. C. (1947). Value and need as organizing factors in perception. *The Journal of Abnormal and Social Psychology, 42,* 33–44.

4. Berns, G. S., Chappelow, J., Zink, C. F., Pagnoni, G., Martin-Skurski, M. E., & Richards, J. (2004). Neurobiological correlates of social conformity and independence during mental rotation. *Biological Psychiatry, 58,* 245–253.

5. Levy, B. (1996). Improving memory in old age through implicit self-stereotyping. *Journal of Personality and Social Psychology, 71,* 1092–1107.

6. Goycoolea, M. V., Goycoolea, H. G., Farfan, C. R., Rodriguez, L. G., Martinez, G. C., & Vidal, R. (1986). Effect of life in industrialized societies on hearing in natives of Easter Island. *The Laryngoscope, 96,* 1391–1396.

7. Holmes, E. R., & Holmes, L. D. (1995). *Other cultures, elder years.*

Thousand Oaks, CA: Sage; Pearson, J. D. (1992). Attitudes and perceptions concerning elderly Samoans in rural Western Samoa, American Samoa, and urban Honolulu. *Journal of Cross-Cultural Gerontology, 7,* 69–88; Thumala, D., Kennedy, B. K., Calvo, E., Gonzalez-Billault, C., Zitko, P., Lillo, P., . . . & Slachevsky, A. (2017). Aging and health policies in Chile: New agendas for research. *Health Systems and Reform, 3,* 253–260.

8. Levy, B. R., Slade, M. D., & Gill, T. (2006). Hearing decline predicted by elders' stereotypes. *The Journals of Gerontology, Series B: Psychological Sciences and Social Sciences, 61,* 82–87.

9. Chasteen, A. L., Pichora-Fuller, M. K., Dupuis, K., Smith, S., & Singh, G. (2015). Do negative views of aging influence memory and auditory performance through self-perceived abilities? *Psychology and Aging, 30,* 881–893.

10. Barber, S. J., & Lee, S. R. (2016). Stereotype threat lowers older adults' self-reported hearing abilities. *Gerontology, 62,* 81–85.

11. Corrigan, P. (2020). Music in an intergenerational key. Next Avenue. https://www.nextavenue.org/music-in-an-intergenerational-key/; Hicks, A. (Director). (2014). *Keep On Keepin' On* [Video]. Retrieved from https://www.amazon.com/Keep-Keepin-Clark-Terry/dp/B00S65TOTE.

12. Parbery-Clark, A., Strait, D. L., Anderson, S., Hittner, E., & Kraus, N. (2011). Musical experience and the aging auditory system: Implications for cognitive abilities and hearing speech in noise. *PLOS ONE, 6,* e18082. https://doi.org/10.1371/journal.pone.0018082; Leopold, W. (2012, January 30). Music training has biological impact on aging process. ScienceDaily. https://www.sciencedaily.com/releases/2012/01/120130172402.htm.

13. Kraus, N., & White-Schwoch, T. (2014). Music training: Lifelong investment to protect the brain from aging and hearing loss. *Acoustics Australia, 42,* 117–123.

14. Ibid.; Kraus, N., & Anderson, S. (2014). Music benefits across the lifespan: Enhanced processing of speech in noise. *Hearing Review, 21,* 18–21; Dr. Nina Kraus on why musical training helps us process the world around us. (2017, May 31). Sound Health. https://medium.com/the-kennedy-center/dr-nina-kraus-on-why-musical-training-helps-us-process-the-world-around-us-6962b42cdf44.

15. Anderson, S., White-Schwoch, T., Parbery-Clark, A., & Kraus, N. (2013). Reversal of age-related neural timing delays with training. *Proceedings of the National Academy of Sciences, 110,* 4357–4362.

16. Burnes, D., Sheppard, C., Henderson, C. R., Wassel, M., Cope, R., Bar-

ber, C., & Pillemer, K. (2019). Interventions to reduce ageism against older adults: A systematic review and meta-analysis. *American Journal of Public Health, 109,* e1–e9.

17. Lively, P. (2013, October 5). So this is old age. *The Guardian.* https://www.theguardian.com/books/2013/oct/05/penelope-lively-old-age.

18. Erikson, J. M. (1988). *Wisdom and the senses.* New York: W. W. Norton, p. 45.

19. Schuster, C., & Carpenter, E. (1996). *Patterns that connect: Social symbolism in ancient and tribal art.* New York: Harry N. Abrams.

20. Simonton, D. K. (1997). Creative productivity: A predictive and explanatory model of career trajectories and landmarks. *Psychological Review, 104,* 66–89.

21. Galenson, D. W. (2010). Late bloomers in the arts and sciences: Answers and questions. National Bureau of Economic Research Working Paper No. w15838, SSRN. https://papers.ssrn.com/sol3/papers.cfm?abstract_id=1578676.

22. Charles, S. T., & Carstensen, L. L. (2010). Social and emotional aging. *Annual Review of Psychology, 61,* 383–409.

23. Steinhardt, A. (1998). *Indivisible by four: Pursuit of harmony.* New York: Farrar, Straus and Giroux.

24. Lindauer, M. S. (2003). *Aging, creativity and art: A positive perspective on late-life development.* New York: Springer.

25. Shahn, B. (1985). *The shape of content.* Cambridge, MA: Harvard University Press.

26. Hathaway, M. (2016, November 30). Harmonious ambition: The resonance of Michelangelo. Virginia Polytechnic Institute and State University. https://vtechworks.lib.vt.edu/handle/10919/74440.

27. Gowing, L. (1966). *Turner: Imagination and reality.* New York: Museum of Modern Art.

28. Spence, J. (1986). *Putting myself in the picture: A political, personal, and photographic autobiography.* London: Camden Press.

29. Pennebaker, J. W., & Stone, L. D. (2003). Words of wisdom: Language use over the life span. *Journal of Personality and Social Psychology, 85,* 291–301.

30. Adams-Price, C. (2017). *Creativity and successful aging: Theoretical and empirical approaches.* New York: Springer.

31. Ibid., pp. 281, 283.

32. Baltes, P. B. (1997). On the incomplete architecture of human ontogeny: Selection, optimization, and compensation as foundation of developmental theory. *American Psychologist, 52,* 366–380; Henahan, D. (1976, March 14). This ageless hero, Rubinstein; He cannot go on like this forever

(though some would not bet on that). In fact, there are now some troubling signs. *The New York Times*. https://www.nytimes.com/1976/03/14/archives/this-ageless-hero-rubinstein-he-cannot-go-on-like-this-forever.html.

33. Grandma Moses is dead at 101; Primitive artist "just wore out." (1961, December 14). *The New York Times*. https://www.nytimes.com/1961/12/14/archives/grandma-moses-is-dead-at-101-primitive-artist-just-wore-out-grandma.html.

34. Henri Matisse (1869–1954). Christies. https://www.christies.com/en/lot/lot-6108776.

35. Ibid.; Museum of Modern Art. (2014). Henri Matisse: The cut-outs. https://www.moma.org/calendar/exhibitions/1429; Murphy, J. (2020, June 9). Henri Matisse: His final years and exhibit. Biography. https://www.biography.com/news/henri-matisse-the-cut-outs-moma.

36. Gardner, H. E. (2011). *Creating minds: An anatomy of creativity seen through the lives of Freud, Einstein, Picasso, Stravinsky, Eliot, Graham, and Gandhi.* New York: Basic Books.

37. Kozbelt, A. (2015). Swan song phenomenon. In S. K. Whitbourne (Ed.), *The encyclopedia of adulthood and aging.* Hoboken, NJ: Wiley.

38. Roth, H. (1997). *From bondage.* London: Picador, p. 188.

39. Claytor, D. (2013, June 26). Retiring in your 30's . . . now what? *Diablo Ballet Blog.* https://diabloballet.wordpress.com/2013/06/26/retiring-in-your-30s-now-what/.

40. To read more about dancer Thomas Dwyer, see Frankel, B. (2011). *What should I do with the rest of my life? True stories of finding success, passion, and new meaning in the second half of life.* New York: Avery.

Chapter 8: Ageism: The Evil Octopus

1. Achenbaum, A. W. (2013). *Robert Butler, MD: Visionary of healthy aging.* New York: Columbia University Press; Bernstein, C. (1969, March 7). Age and race fears seen in housing opposition. *Washington Post.*

2. Butler, R. N. (1975). *Why survive? Being old in America.* New York: Harper & Row, p. 12.

3. Ober Allen, J., Solway, E., Kirch, M., Singer, D., Kullgren, J., & Malani, P. (2020, July). Everyday ageism and health. National Poll on Healthy Aging. University of Michigan. http://hdl.handle.net/2027.42/156038.

4. World Health Organization. (2021). *Global Report on Ageism.* Geneva: World Health Organization. https://www.who.int/teams/social-determinants-of-health/demographic-change-and-healthy-ageing/combating-ageism/global-report-on-ageism.

5. International Longevity Center. Anti-Ageism Taskforce. (2006). *Ageism in America*. New York: International Longevity Center-USA.

6. Stratton, C., Andersen, L., Proulx, L., & Sirotich, E. (2021). When apathy is deadlier than COVID-19. *Nature Aging, 1*, 144–145.

7. Ng, R., Allore, H. G., Trentalange, M., Monin, J. K., & Levy, B. R. (2015). Increasing negativity of age stereotypes across 200 years: Evidence from a database of 400 million words. *PLOS ONE, 10*, e0117086.

8. Levy, B. R. (2009). Stereotype embodiment: A psychosocial approach to aging. *Current Directions in Psychological Science, 18*, 332–336.

9. Levy, B. R., and Banaji, M. R. (2004). Implicit ageism. In T. Nelson (Ed.), *Ageism: Stereotyping and prejudice against older persons* (pp. 49–75). Cambridge, MA: MIT Press.

10. Estes, C., Harrington, C., & Pellow, D. (2001). The medical-industrial complex and the aging enterprise. In C. L. Estes, *Social policy & aging: A critical perspective* (pp. 165–186). Thousand Oaks, CA: Sage; Beauty Packaging Staff. (2020). Anti-aging market forecasted to surpass $421.4 billion in revenue by 2030. Beauty Packaging. https://www.beauty packaging.com/contents/view_breaking-news/2020-09-23/anti-aging -market-forecasted-to-surpass-4214-billion-in-revenue-by-2030; Guttmann, A. (2020, November 23). Social network advertising revenues in the United States from 2017 to 2021. Statista. https://www.statista.com /statistics/271259/advertising-revenue-of-social-networks-in-the-us/; Guttmann, A. (2021, February 4). Estimated aggregate revenue of U.S. advertising, public relations, and related service industry from 2004 to 2020. Statista. https://www.statista.com/statistics/183932/estimated -revenue-in-advertising-and-related-services-since-2000/; Grand View Research. (2020, March). U.S. long term care market size, share and trends analysis by service (home healthcare, hospice, nursing care, assisted living facilities), and segment forecasts. https://www.grandview research.com/industry-analysis/us-long-term-care-ltc-market.

11. McGuire, S. L. (2016). Early children's literature and aging. *Creative Education, 7*, 2604–2612.

12. Gilbert, C. N., & Ricketts, K. G. (2008). Children's attitudes toward older adults and aging: A synthesis of research. *Educational Gerontology, 34*, 570–586.

13. Seefeldt, C., Jantz, R. K., Galper, A., & Serock, S. (1977). Using pictures to explore children's attitudes toward the elderly. *The Gerontologist, 17*, 506–512.

14. Middlecamp, M., & Gross, D. (2002). Intergenerational daycare and pre-

schoolers' attitudes about aging. *Educational Gerontology, 28,* 271–288; Kwong See, S. T., Rasmussen, C., & Pertman, S. Q. (2012). Measuring children's age stereotyping using a modified Piagetian conservation task. *Educational Gerontology, 38,* 149–165; Seefeldt, C., Jantz, R., Galper, A., & Serock, K. (1977). Children's attitudes toward the elderly: Educational implications. *Educational Gerontology, 2,* 301–310.

15. Levy, B. R. (2009). Stereotype embodiment: A psychosocial approach to aging. *Current Directions in Psychological Science, 18,* 332–336.

16. Vitale-Aussem, J. (2018, September 11). "Dress like a 100-year old" day: A call to action. ChangingAging. https://changingaging.org/ageism /dress-like-a-100-year-old-day-a-call-to-action/.

17. Levy, B. R., Zonderman, A. B., Slade, M. D., & Ferrucci, L. (2009). Age stereotypes held earlier in life predict cardiovascular events in later life. *Psychological Science, 20,* 296–298.

18. Ridder, M. (2021, January 27). Value of the global anti-aging market 2020–2026. Statista. Retrieved July 13, 2021, from https://www.statista .com/statistics/509679/value-of-the-global-anti-aging-market/.

19. Diller, V. (2011, November 17). Too young to look old? Dealing with fear of aging. *HuffPost.* https://www.huffpost.com/entry/aging-fear _b_812792?; Kilkenny, K. (2017, August 30). How anti-aging cosmetics took over the beauty world. *Pacific Standard.* https://psmag.com/social -justice/how-anti-aging-cosmetics-took-over-the-beauty-world.

20. Blanchette, A. (2017, January 28). Botox is booming among millennials— some as young as 18. *Star Tribune.* https://www.startribune.com/botox- is-booming-among-millennials-some-as-young-as-18/412049303/.

21. North, A. (2021, June 15). Free the wrinkle: The pandemic could help Americans finally embrace aging skin. Vox. https://www.vox .com/22526590/wrinkles-skin-botox-aging-pandemic-filler.

22. Schiffer, J. (2021, April 8). How barely-there Botox became the norm. *The New York Times.* https://www.nytimes.com/2021/04/08/style/self -care-how-barely-there-botox-became-the-norm.html.

23. Market Insider. (2019, November 28). Every thirteenth man has a hair transplant according to Bookimed study. https://markets.businessin sider.com/news/stocks/every-13th-man-has-a-hair-transplant-according -to-bookimed-study-1028724168.

24. *Vermont Country Store Catalogue,* 2015, p. 21.

25. Calasanti, T., Sorensen, A., & King, N. (2012). Anti-ageing advertisements and perceptions of ageing. In V. Ylänne (Ed.), *Representing ageing.* London: Palgrave Macmillan.

26. *Smithsonian.* (2016). *Choose life: Grow young with HGH.* 47, p. 105.

27. Clayton, P., Banerjee, I., Murray, P., & Renehan, A. G. (2011). Growth hormone, the insulin-like growth factor axis, insulin and cancer risk. *Nature Reviews Endocrinology, 7,* 11–24.

28. Cornell, E. M., Janetos, T. M., & Xu, S. (2019). Time for a makeover-cosmetics regulation in the United States. *Journal of Cosmetic Dermatology, 18,* 2040–2047. https://onlinelibrary.wiley.com/doi/full/10.1111/jocd.12886; Mehlman, M. J., Binstock, R. H., Juengst, E. T., Ponsaran, R. S., & Whitehouse, P. J. (2004). Anti-aging medicine: Can consumers be better protected? *The Gerontologist, 44,* 304–310.

29. Perls, T. T. (2004). Anti-aging quackery: Human growth hormone and tricks of the trade—More dangerous than ever. *The Journals of Gerontology, Series A: Biological Sciences and Medical Sciences, 59,* 682–691.

30. Lieberman, T. (2013, March 14). The enduring myth of the greedy geezer. *Columbia Journalism Review.* https://archives.cjr.org/united_states_project/the_enduring_myth_of_the_greed.php.

31. Levy, B. R., & Schlesinger, M. (2005). When self-interest and age stereotypes collide: Elders' preferring reduced funds for programs benefiting themselves. *Journal of Aging and Social Policy, 17,* 25–39.

32. Frumkin, H., Fried, L., & Moody, R. (2012). Aging, climate change, and legacy thinking. *American Journal of Public Health, 102,* 1434–1438; Konrath, S., Fuhrel-Forbis, A., Lou, A., & Brown, S. (2012). Motives for volunteering are associated with mortality risk in older adults. *Healthy Psychology, 31,* 87–96; Benefactor Group. Sixty and over: Elders and philanthropic investments. https://benefactorgroup.com/sixty-and-over-elders-and-philanthropic-investments/; Soergel, A. (2019, November 18). California, Texas caregivers offer billions in free care. *US News & World Report.* https://www.usnews.com/news/best-states/articles/2019-11-18/family-caregivers-in-us-provide-470-billion-of-unpaid-care.

33. Robinson, J. D., & Skill, T. (2009). The invisible generation: Portrayals of the elderly on prime-time television. *Communication Reports, 8,* 111–119. https://doi.org/10.1080/08934219509367617; Zebrowitz, L. A., & Montepare, J. M. (2000). "Too young, too old": Stigmatizing adolescents and elders. In T. F. Heatherton, R. E. Kleck, M. R. Hebl, & J. G. Hull (Eds.), *The social psychology of stigma* (pp. 334–373). New York: Guilford Press.

34. Follows, S. (2015, September 7). How old are Hollywood screenwriters? Stephen Follows Film Data and Education. https://stephenfollows.com/how-old-are-hollywood-screenwriters/.

35. Geena Davis Institute on Gender in Media. (2018). The reel truth:

Women aren't seen or heard. https://seejane.org/research-informs-em powers/data/.

36. Smith, S., Pieper, K., & Chouiti, M. (2018). Still rare, still ridiculed: Portrayals of senior characters on screen: Popular films from 2015 and 2016. USC Annenberg School for Communication and Journalism. http://assets.uscannenberg.org/docs/still-rare-still-ridiculed.pdf.

37. Sperling, N. (2020, September 8). Academy explains diversity rules for best picture Oscar. *The New York Times*. https://www.nytimes.com/2020 /09/08/movies/oscars-diversity-rules-best-picture.html?.

38. Ibid.

39. She won one for acting and one for fighting for increasing the inclusion of women on-screen.

40. Geena Davis Institute on Gender in Media. (2018). The reel truth: Women aren't seen or heard. https://seejane.org/research-informs-em powers/data/.

41. Newsdesk. Geena Davis disheartened by Hollywood attitudes to age and gender. (2020, August 11). *Film News*. https://www.film-news.co.uk /news/UK/78030/Geena-Davis-disheartened-by-Hollywood-attitudes -to-age-and-gender.

42. Smith, N. (2020, October 30). Geena Davis reacts to the "dismal" findings of her center's study on ageism in Hollywood: "It's a shame." *People*. https://people.com/movies/geena-davis-reacts-to-the-dismal-findings-of -her-centers-study-on-ageism-in-hollywood/.

43. Donlon, M., & Levy, B. R. (2005). Re-vision of older television characters: Stereotype-awareness intervention. *Journal of Social Issues, 61*, 307–319.

44. Gerbner, G., Gross, L., Signorielli, N., & Morgan, M. (1980). Aging with television: Images in television drama and conceptions of social reality. *Journal of Communication, 30*, 37–47; Harwood, J., & Anderson, K. (2002). The presence and portrayal of social groups on prime-time television. *Communication Reports, 15*, 81–97.

45. Safronova, V., Nikas, J., & Osipova, N. (2017, September 5). What it's truly like to be a fashion model. *The New York Times*. https://www.ny times.com/2017/09/05/fashion/models-racism-sexual-harassment-body -issues-new-york-fashion-week.html.

46. *The New York Times* video accompanying article, Safronova, V., Nikas, J., Osipova, N. (2017, September 5). What it's truly like to be a fashion model. *The New York Times*. https://www.nytimes.com/2017/09/05 /fashion/models-racism-sexual-harassment-body-issues-new-york-fash ion-week.html.

47. Gillin, J. (2017, October 4). The more outrageous, the better: How clickbait ads make money for fake news sites. PolitiFact. https://www.politifact.com/article/2017/oct/04/more-outrageous-better-how-clickbait-ads-make-mone/.

48. Zulli, D. (2018). Capitalizing on the look: Insights into the glance, attention economy, and Instagram. *Critical Studies in Media Communication, 35,* 137–150.

49. Levy, B. R., Chung, P. H., Bedford, T., & Navrazhina, K. (2014). Facebook as a site for negative age stereotypes. *Gerontologist, 54,* 172–176.

50. Facebook Community Standards. Objectionable content: Hate speech. Retrieved March 14, 2021, from https://www.facebook.com/communitystandards/.

51. Jimenez-Sotomayor, M. R., Gomez-Moreno, C., & Soto-Perez-de-Celis, E. (2020). Coronavirus, ageism, and Twitter: An evaluation of tweets about older adults and COVID-19. *Journal of the American Geriatrics Society, 68,* 1661–1665.

52. Oscar, N., Fox, P. A., Croucher, R., Wernick, R., Keune, J., & Hooker, K. (2017). Machine learning, sentiment analysis, and tweets: An examination of Alzheimer's disease stigma on Twitter. *The Journals of Gerontology, Series B: Psychological Sciences and Social Sciences, 72,* 742–751.

53. Gabbatt, A. (2019, March 28). Facebook charged with housing discrimination in targeted ads. *The Guardian.* https://www.theguardian.com/technology/2019/mar/28/facebook-ads-housing-discrimination-charges-us-government-hud.

54. The Associated Press (2020, July 1). Lawsuit accuses property managers of ageist ads. Finance & Commerce. https://finance-commerce.com/2020/07/lawsuit-accuses-property-managers-of-ageist-ads/.

55. Terrell, K. (2019, March 20). Facebook reaches settlement in age discrimination lawsuits. AARP. https://www.aarp.org/work/working-at-50-plus/info-2019/facebook-settles-discrimination-lawsuits.html; Kofman, A., & Tobin, A. (2019, December 13). Facebook ads can still discriminate against women and older workers, despite a civil rights settlement. ProPublica. https://www.propublica.org/article/facebook-ads-can-still-discriminate-against-women-and-older-workers-despite-a-civil-rights-settlement.

56. Pelisson, A., & Hartmans, A. (2017, September 11). The average age of employees at all the top tech companies, in one chart. *Insider.* https://www.businessinsider.com/median-tech-employee-age-chart-2017-8.

57. Freedman, M., & Stamp, T. (2018, June 6). The US isn't just getting

older. It's getting more segregated by age. *Harvard Business Review.* https://hbr.org/2018/06/the-u-s-isnt-just-getting-older-its-getting-more -segregated-by-age.

58. Ruggles, S., & Brower, S. (2003). The measurement of family and household composition in the United States, 1850–1999. *Population and Development Review, 29,* 73–101.

59. Winkler, R. (2013). Segregated by age: Are we becoming more divided? *Population Research and Policy Review, 32,* 717–727.

60. Intergenerational Foundation. (2016). "Generations apart? The growth of age segregation in England and Wales." https://www.if.org.uk /research-posts/generations-apart-the-growth-of-age-segregation-in -england-and-wales/.

61. Hagestad, G. O., & Uhlenberg, P. (2005). The social separation of old and young: A root of ageism. *Journal of Social Issues, 61,* 343–360.

62. Kelley, O. (2020, October 8). This man was fired due to ageism and being "too American." Ladders. https://www.theladders.com/career -advice/this-man-was-fired-due-to-ageism-and-being-too-american.

63. Gosselin, P. (2018, December 28). If you're over 50, chances are the decision to leave a job won't be yours. ProPublica. https://www.propublica.org /article/older-workers-united-states-pushed-out-of-work-forced-retirement.

64. FitzPatrick, C. S. (2014). Fact sheet: Age discrimination. AARP Office of Policy Integration. https://www.aarp.org/ppi/info-2015/age-discrim ination.html.

65. Gosselin, P. (2017). Supreme Court won't take up R. J. Reynolds age discrimination case. ProPublica. https://www.propublica.org/article /supreme-court-rj-reynolds-age-discrimination-case.

66. Halbach, J. H., & Haverstock, P. M. (2009). Supreme Court sets higher burden for plaintiffs in age discrimination claims. Larkin Hoffman. https://www.larkinhoffman.com/media/supreme-court-sets-higher-bur den-for-plaintiffs-in-age-discrimination-claims.

67. Age Smart Employer: Columbia Aging Center. (2021, May 24). The advantages of older workers. https://www.publichealth.columbia.edu /research/age-smart-employer/advantages-older-workers; Raymo, J. M., Warren, J. R., Sweeney, M. M., Hauser, R. M., & Ho, J. H. (2010). Later-life employment preferences and outcomes: The role of mid-life work experiences. *Research on Aging, 32,* 419–466.

68. Rosen, W. (2017, May 16). How the first broad-spectrum antibiotic emerged from Missouri dirt. *Popular Science.* https://www.popsci.com /how-first-broad-spectrum-antibiotic-emerged-from-dirt/.

69. Butler, R. N. (2008). *The longevity revolution: The benefits and challenges of living a long life.* New York: PublicAffairs.

70. Chang, E., Kannoth, S., Levy, S., Wang, S., Lee, J. E., & Levy, B. R. (2020). Global reach of ageism on older persons' health: A systematic review. *PLOS ONE, 15,* e0220857. https://doi.org/10.1371/journal.pone.0220857.

71. Loch, C., Sting, F., Bauer, N., & Mauermann, H. (2010). The globe: How BMW is defusing the demographic time bomb. *Harvard Business Review.* https://hbr.org/2010/03/the-globe-how-bmw-is-defusing-the-demographic-time-bomb.

72. Conley, C. (2018). *Wisdom at work: The making of a modern elder.* New York: Currency, p. 117.

73. Levy, B. R. (2009). Stereotype embodiment: A psychosocial approach to aging. *Current Directions in Psychological Science, 18,* 332–336; Estes, C. L., & Binney, E. A. (1989). The biomedicalization of aging—dangers and dilemmas. *The Gerontologist, 29,* 587–596.

74. Estes, C., Harrington, C., & Pellow, D. (2001). The medical-industrial complex and the aging enterprise. In C. L. Estes, *Social policy & aging: A critical perspective* (pp. 165–186). Thousand Oaks, CA: Sage.

75. Makris, U. E., Higashi, R. T., Marks, E. G., Fraenkel, L., Sale, J. E., Gill, T. M., & Reid, M. C. (2015). Ageism, negative attitudes, and competing co-morbidities—why older adults may not seek care for restricting back pain: A qualitative study. *BMC Geriatrics, 15,* 39.

76. Ibid.

77. Aronson, L. (2019). *Elderhood: Redefining aging, transforming medicine, reimagining life.* New York: Bloomsbury.

78. Centers for Disease Control and Prevention. (2020, May). HIV Surveillance Report, 2018 (Updated). http://www.cdc.gov/hiv/library/reports/hiv-surveillance.html.

79. Butler, R. (1989). Dispelling ageism: The cross-cutting intervention. *The Annals of the American Academy of Political and Social Science, 503,* 138–147.

80. Meiboom, A. A., de Vries, H., Hertogh, C. M., & Scheele, F. (2015). Why medical students do not choose a career in geriatrics: A systematic review. *BMC Medical Education, 15,* 101.

81. Remmes, K., & Levy, B. R. (2005). Medical school training and ageism. In E. B. Palmore, L. Branch, & D. Harris (Eds.), *Encyclopedia of ageism.* Philadelphia: Routledge.

82. Cayton, H. (2006). The alienating language of health care. *Journal of the Royal Society of Medicine, 99,* 484.

83. Achenbaum, A. W. (2013). *Robert Butler, MD: Visionary of healthy aging.* New York: Columbia University Press, p. 84.

84. Hudson, J., Waters, T., Holmes, M., Agris, S., Seymour, D., Thomas, L., & Oliver, E. J. (2019). Using virtual experiences of older age: Exploring pedagogical and psychological experiences of students. *VRAR, 18,* 61–72.

85. There are ways of inducing empathy without risking the enhancement of negative age beliefs. For instance, a study found that asking participants to write an essay from the perspective of an older person decreased negative age stereotypes. Galinsky, A. D., & Moskowitz, G. B. (2000). Perspective-taking: Decreasing stereotype expression, stereotype accessibility, and in-group favoritism. *Journal of Personality and Social Psychology, 78,* 708–724.

86. Meiboom, A. A., de Vries, H., Hertogh, C. M., & Scheele, F. (2015). Why medical students do not choose a career in geriatrics: A systematic review. *BMC Medical Education, 15,* 101.

87. Siu, A., & Beck, J. C. (1990). Physician satisfaction with career choices in geriatrics. *The Gerontologist, 30,* 529–534.

88. Wyman, M. F., Shiovitz-Ezra, S., & Bengel, J. (2018). Ageism in the health care system: Providers, patients, and systems. In L. Ayalon & C. Tesch-Römer (Eds.), *Contemporary perspectives on ageism* (pp. 193–212). New York: Springer.

89. Chang, E., Kannoth, S., Levy, S., Wang, S., Lee, J. E., & Levy, B. R. (2020). Global reach of ageism on older persons' health: A systematic review. *PLOS ONE, 15*(1): e0220857. https://doi.org/10.1371/journal.pone.0220857.

90. The method involved identifying the excess cost due to ageism (which included both age discrimination and negative age beliefs) that went beyond the routine costs of the illnesses themselves. Levy, B. R., Slade, M., Chang, E. S., Kannoth, S., & Wang, S. H. (2020). Ageism amplifies cost and prevalence of health conditions. *The Gerontologist, 60,* 174–181.

91. Ibid.

92. Kim, D. D., & Basu, A. (2016). Estimating the medical care costs of obesity in the United States: Systematic review, meta-analysis, and empirical analysis. *Value in Health, 19,* 602–613; Tsai, A. G., Williamson, D. F., & Glick, H. A. (2011). Direct medical cost of overweight and obesity in the USA: A quantitative systematic review. *Obesity Reviews, 12,* 50–61.

93. Cedars-Sinai. (2021, January 18). LeVar Burton Hosts Cedars-Sinai

Celebration of Martin Luther King Jr. https://www.cedars-sinai.org/newsroom/levar-burton-hosts-cedars-sinai-celebration-of-martin-luther-king-jr/.

94. Epel, E. S., Crosswell, A. D., Mayer, S. E., Prather, A. A., Slavich, G. M., Puterman, E., & Mendes, W. B. (2018). More than a feeling: A unified view of stress measurement for population science. *Frontiers in Neuroendocrinology, 49,* 146–169; McEwen, B. S. (2013). The brain on stress: Toward an integrative approach to brain, body, and behavior. *Perspectives on Psychological Science, 8,* 673–675.

95. Butler, R. N. (2008). *The longevity revolution: The benefits and challenges of living a long life.* New York: PublicAffairs; Butler, R. N. (1975). *Why survive? Being old in America.* New York: Harper & Row.

96. Bekiempis, V. (2021, February 20). "Alarming surge" in anti-Asian violence across US terrifies community members. *The Guardian.* https://www.theguardian.com/us-news/2021/feb/20/anti-asian-violence-us-bigotry.

97. Kim, J., & McCullough, R. (2019). When it comes to aging, intersectionality matters. Caring Across Generations. https://caringacross.org/when-it-comes-to-aging-intersectionality-matters/; Creamer, J. (2020, September 15). Census data shows inequalities persist despite decline in poverty for all major race and Hispanic origin groups. *Lake County News.* https://www.lakeconews.com/news/66713-census-data-shows-inequalities-persist-despite-decline-in-poverty-for-all-major-race-and-hispanic-origin-groups.

98. Kaelber, L. A. (2012). The invisible elder: The plight of the elder Native American. *Marquette Elder's Advisor, 3,* 46–57; Ellis, R. (2021, February 5). COVID deadlier for Native Americans than other groups. WebMD. https://www.webmd.com/lung/news/20210204/covid-deadlier-for-native-americans-than-other-groups.

Chapter 9: Individual Age Liberation: How to Free Your Mind

1. Marottoli, R. A., & Coughlin, J. F. (2011). Walking the tightrope: Developing a systems approach to balance safety and mobility for an aging society. *Journal of Aging & Social Policy, 23,* 372–383; Tortorello, M. (2017, June 1). How seniors are driving safer, driving longer. *Consumer Reports.* https://www.consumerreports.org/elderly-driving/how-seniors-are-driving-safer-driving-longer/; Leefeldt, E. & Danise, A. (2021, March 16). Senior drivers are safer than previously thought. *Forbes.* https://www.forbes.com/advisor/car-insurance/seniors-driving

-safer/; American Occupational Therapy Association. Myths and realities about older drivers. https://www.aota.org/Practice/Productive-Aging/Driving/Clients/Concern/Myths.aspx.

2. Williams, K., Kemper, S., & Hummert, M. L. (2005). Enhancing communication with older adults: Overcoming elderspeak. *Journal of Psychosocial Nursing and Mental Health Services, 43,* 12–16.

3. Corwin, A. I. (2018). Overcoming elderspeak: A qualitative study of three alternatives. *The Gerontologist, 58,* 724–729.

4. Levy, B. R., Pilver, C., Chung, P. H., & Slade, M. D. (2014). Subliminal strengthening: Improving older individuals' physical function over time with an implicit-age-stereotype intervention. *Psychological Science, 25,* 2127–2135.

5. Ferro, S. (2018, April 18). The "Scully effect" is real: Female *X-Files* fans more likely to go into STEM. Mental Floss. https://www.mentalfloss.com/article/540530/scully-effect-female-x-files-viewers-stem-careers.

6. Levy, B. R., Pilver, C., Chung, P. H., & Slade, M. D. (2014). Subliminal strengthening: Improving older individuals' physical function over time with an implicit-age-stereotype intervention. *Psychological Science, 25,* 2127–2135.

7. Langer, E. J. (2009). *Counter clockwise: Mindful health and the power of possibility.* New York: Ballantine Books.

8. Dasgupta, N., & Greenwald, A. G. (2001). On the malleability of automatic attitudes: Combating automatic prejudice with images of admired and disliked individuals. *Journal of Personality and Social Psychology, 81,* 800–814.

9. Fung, H. H., Li, T., Zhang, X., Sit, I. M. I., Cheng, S., & Isaacowitz, D. M. (2015). Positive portrayals of old age do not always have positive consequences. *The Journals of Gerontology, Series B: Psychological Sciences and Social Sciences, 70,* 913–924.

10. Lowsky, D. J., Olshansky, S. J., Bhattacharya, J., & Goldman, D. P. (2014). Heterogeneity in healthy aging. *The Journals of Gerontology, Series A: Biological Sciences and Medical Sciences, 69,* 640–649.

11. Applewhite, A. (2016). *This chair rocks: A manifesto against ageism.* Networked Books.

12. Plaut, V. C., Thomas, K. M., Hurd, K., & Romano, C. A. (2018). Do color blindness and multiculturalism remedy or foster discrimination and racism? *Current Directions in Psychological Science, 27,* 200–206.

13. Krajeski, J. (2008, September 19). This is water. *The New Yorker.* https://www.newyorker.com/books/page-turner/this-is-water.

14. Zwirky, A. (2017, June 14). There's a name for that: The Baader-Meinhof phenomenon: When a thing you just found out about suddenly seems to crop up everywhere. *Pacific Standard.* https://psmag.com/social-justice /theres-a-name-for-that-the-baader-meinhof-phenomenon-59670.

15. Burnes, D., Sheppard, C., Henderson, C. R., Wassel, M., Cope, R., Barber, C., & Pillemer, K. (2019). Interventions to reduce ageism against older adults: A systematic review and meta-analysis. *American Journal of Public Health, 109,* e1–e9. https://doi.org/10.2105/ajph.2019.305123.

16. Ross, L. (1977). The intuitive psychologist and his shortcomings: Distortions in the attribution process. In L. Berkowitz (Ed.), *Advances in experimental social psychology* (pp. 173–220). New York: Academic Press.

17. Skurnik, I., Yoon, C., Park, D. C., & Schwarz, N. (2005). How warnings about false claims become recommendations. *Journal of Consumer Research, 31,* 713–724; Tucker, J., Klein, D., & Elliott, M. (2004). Social control of health behaviors: A comparison of young, middle-aged, and older adults. *The Journals of Gerontology, Series B: Psychological Sciences and Social Sciences, 59,* 147–150; Cotter, K. A. (2012). Health-related social control over physical activity: Interactions with age and sex. *Journal of Aging Research.* https://doi.org/10.1155/2012/321098.

18. National Academies of Sciences, Engineering, and Medicine. (2019). *Integrating social care into the delivery of health care: Moving upstream to improve the nation's health.* Washington, DC: The National Academies Press. https://doi.org/10.17226/25467.

19. Levy, B. R., Chung, P. H., Slade, M. D., Van Ness, P. H., & Pietrzak, R. H. (2019). Active coping shields against negative aging self-stereotypes contributing to psychiatric conditions. *Social Science and Medicine, 228,* 25–29.

20. Ashton, A. (2021, February 22). Anonymous asked: What are your thoughts on the phrase "for the young at heart"? *Yo, is this ageist?* https://yoisthisageist.com/post/643858558300670712/what-are-your -thoughts-on-the-phrase-for-the.

21. *Vogue.* (2019, May 3). Madonna on motherhood and fighting ageism: "I'm being punished for turning 60." https://www.vogue.co.uk/article /madonna-on-ageing-and-motherhood.

22. De Souza, A. (2015, September 23). Ageism exists in Hollywood, says Robert De Niro. *The Straits Times.* https://www.straitstimes.com /lifestyle/entertainment/ageism-exists-in-hollywood-says-robert-de -niro-72.

23. Hsu, T. (2019, September 23). Older people are ignored and distorted

in ageist marketing, report finds. *The New York Times*. https://www.ny times.com/2019/09/23/business/ageism-advertising-aarp.html.

24. Dan, A. (2016, September 13). Is ageism the ugliest "ism" on Madison Ave? *Forbes*. https://www.forbes.com/sites/avidan/2016/09/13/is-ageism -the-ugliest-ism-on-madison-avenue/?sh=695108ae557c.

25. Ad Council. (2015, March 3). Love has no labels. Diversity and Inclusion. [Video]. https://www.youtube.com/watch?v=PnDgZuGIhHs.

Chapter 10: Societal Age Liberation: A New Social Movement

1. Dychtwald, K. (2012, May 31). Remembering Maggie Kuhn: Gray Panthers founder on the 5 myths of aging. *HuffPost*. http://www.huffing tonpost.com/ken-dychtwald/the-myths-of-aging_b_1556481.html.

2. Douglas, S. (2020, September 8). The forgotten history of the radical "elders of the tribe." *The New York Times*. https://www.nytimes .com/2020/09/08/opinion/sunday/gray-panthers-maggie-kuhn.html.

3. Fokart, B. (1995, April 23). Maggie Kuhn, 89; Iconoclastic founder of Gray Panthers. *Los Angeles Times*. https://www.latimes.com/archives /la-xpm-1995-04-23-mn-58042-story.html; La Jeunesse, M. (2019, August 2). Who was Maggie Kuhn, co-founder of the elder activist group the Gray Panthers? *Teen Vogue*. https://www.teenvogue.com/story/who -was-maggie-kuhn-co-founder-elder-activist-group-gray-panthers.

4. La Jeunesse, M. (2019, August 2). Who was Maggie Kuhn, co-founder of the elder activist group the Gray Panthers? *Teen Vogue*. https://www .teenvogue.com/story/who-was-maggie-kuhn-co-founder-elder-activist -group-gray-panthers.

5. Butler, R. N. (1975). *Why survive? Being old in America*. New York: Harper & Row, p. 341.

6. Mapes, J. (2019, January 10). Senator Ron Wyden, who started career as senior advocate, is now 65. *The Oregonian*. https://www.oregonlive.com /mapes/2014/05/sen_ron_wyden_who_started_care_1.html.

7. Pew Research Center. (2019, May 14). Attitudes on same-sex marriage. https://www.pewforum.org/fact-sheet/changing-attitudes-on-gay -marriage/; McCarthy, J. (2019). U.S. support for gay marriage stable, at 63%. Gallup. https://news.gallup.com/poll/257705/support-gay -marriage-stable.aspx.

8. Morris, A. D. (1984). *The origins of the civil rights movement: Black communities organizing for change*. New York: Free Press; Levy, B. R. (2017). Age-stereotype paradox: A need and opportunity for social change. *The Gerontologist, 57,* 118–126.

9. McAdam, D. (1982). *Political process and the development of the Black insurgency, 1930–1970.* Chicago: The University of Chicago Press.

10. Morris, A. D. (1984). *The origins of the civil rights movement: Black communities organizing for change.* New York: Free Press.

11. Levy, B. R. (2017). Age-stereotype paradox: Opportunity for social change. *Gerontologist, 57,* 118–126.

12. Theatre of the Oppressed NYC. (2018, April 24). April 12 Recap: The Runaround. https://www.tonyc.nyc/therunaroundrecap.

13. Comedy Central. (2015, April 22). *Inside Amy Schumer.* Last F**kable Day. [Video]. YouTube. https://www.youtube.com/watch?v=XPpsI8mWKmg.

14. Bunis, D. (2018, April 30). The immense power of the older voter. AARP. https://www.aarp.org/politics-society/government-elections/info-2018/power-role-older-voters.html.

15. Thompson, L. E., Barnett, J. R., & Pearce, J. R. (2009). Scared straight? Fear-appeal anti-smoking campaigns, risk, self-efficacy and addiction. *Health, Risk & Society, 11,* 181–196; van Reek, J., & Adriaanse, H. (1986). Anti-smoking information and changes of smoking behaviour in the Netherlands, UK, USA, Canada and Australia. In D. S. Leathar, G. B. Hastings, K. O'Reilly, & J. K. Davies (Eds.), *Health education and the media II* (pp. 45–50). Oxford: Pergamon.

16. Best, W. (2018, December 17). Gray is the new black: Baby boomers still outspend millennials. Visa. https://usa.visa.com/partner-with-us/visa-consulting-analytics/baby-boomers-still-outspend-millennials.html.

17. International Longevity Centre-UK. (2019, December 5). "Neglected": Opportunities of ageing could add 2% to UK GDP. Global Coalition on Aging. https://globalcoalitiononaging.com/2019/12/05/neglected-opportunities-of-ageing-could-add-2-to-uk-gdp/.

18. Robinson, T., Callister, M., Magoffin, D., & Moore, J. (2007). The portrayal of older characters in Disney animated films. *Journal of Aging Studies, 21,* 203–213; Kessler, E., Rakoczy, K., & Staudinger, U. M. (2004). The portrayal of older people in prime time television series: The match with gerontological evidence. *Ageing and Society, 24,* 531–552.

19. Levy, B. R., Chung, P. H., Bedford, T., & Navrazhina, K. (2014). Facebook as a site for negative age stereotypes. *Gerontologist, 54,* 172–176.

20. Haasch, P. (2020, September 16). All the celebrities protesting Facebook and Instagram by pausing social posts on Stop Hate for Profit Day. *Insider.* https://www.insider.com/celebrities-kim-kardashian-stop-hate-for-profit-instagram-facebook-boycott-protest-2020; Stop Hate for Profit. https://www.stophateforprofit.org; Frenkel, S. (2020, Octo-

ber 20). Facebook bans content about Holocaust denial from its site. *The New York Times*. https://www.nytimes.com/2020/10/12/technology/facebook-bans-holocaust-denial-content.html.

21. Levy, B. R., Slade, M., Chang, E. S., Kannoth, S., & Wang, S. H. (2020). Ageism amplifies cost and prevalence of health conditions. *The Gerontologist, 60,* 174–181.

22. Li, Z., & Dalaker, J. (2021, April 14). Poverty among the population aged 65 and older. Congressional Research Service. https://fas.org/sgp/crs/misc/R45791.pdf.

23. Charlton, J. I. (2000). *Nothing about us without us: Disability oppression and empowerment.* Berkeley: University of California Press.

24. Calvario, L. (2019, April 18). Reese Witherspoon proudly embraces her fine lines and gray hair. *ET.* https://www.etonline.com/reese-witherspoon-proudly-embraces-her-fine-lines-and-gray-hair-123665.

25. Names and details have been changed to protect Wrinkle Salon participants' privacy.

26. Kendi, I. X. (2019). *How to be an antiracist.* New York: One World, pp. 113–114.

27. Now This. (2019, May 29). At 67 years old, JoAni Johnson is the new face of Rihanna's FENTY fashion line. https://nowthisnews.com/videos/her/joani-johnson-is-the-new-face-of-rihannas-fenty-fashion-line; Hicklin, A. (2019, September 15). JoAni Johnson: The sexagenarian model defying convention. *The Guardian.* https://www.theguardian.com/fashion/2019/sep/15/joani-johnson-model-fenty-career-began-at-65-ageing-interview; Foussianes, C. (2019, May 28). Rihanna casts JoAni Johnson, a stunning 68-year-old model, in her Fenty campaign. *Town & Country.* https://www.townandcountrymag.com/style/fashion-trends/a27610015/rihanna-joani-johnson-fenty-older-model/; Coley, P. (2020, January 12). JoAni Johnson: 67-year-old model personally picked by Rihanna for Fenty. *Spectacular Magazine.* https://spectacularmag.com/2020/01/12/joani-johnson-67-year-old-model-personally-picked-by-rihanna-for-fenty/.

28. Lewis, D. C., Desiree, K., & Seponski, D. M. (2011). Awakening to the desires of older women: Deconstructing ageism within fashion magazines. *Journal of Aging Studies, 25,* 101–109.

29. Ewing, A. S. (2019, May 27). Senior slay: JoAni Johnson, 68, proves ageless beauty, grace, and power. The Root. https://theglowup.theroot.com/senior-slay-joani-johnson-68-proves-ageless-beauty-1835048108.

30. World Health Organization. (2021). Combatting ageism. https://www

.who.int/teams/social-determinants-of-health/demographic-change
-and-healthy-ageing/combatting-ageism.

31. Centola, D., Becker, J., Brackbill, D., & Baronchelli, A. (2018). Experimental evidence for tipping points in social convention. *Science, 360*, 1116–1119.

32. Szmigiera, M. (2021, March 30). Distribution of the global population in 2020, by age group and world region. Statista. https://www.statista .com/statistics/875605/percentage-share-of-world-population-by-age -and-by-world-region/.

Afterword: A Town Free of Ageism

1. The description of Greensboro is based on individual interviews I conducted with its residents. In order to convey a sense of the interconnectedness that is a prominent and admirable feature of the town, I have merged these interviews and provided them with a Greensboro context.

2. Beach, B., & Bamford, S. M. (2014). *Isolation: The emerging crisis for older men. A report exploring experiences of social isolation and loneliness among older men in England.* London: International Longevity Center-UK.

3. Greensboro Historical Society. (1990). *The history of Greensboro: The first two hundred years.* Greensboro, VT: Greensboro Historical Society.

Appendix 1: ABC Method to Bolster Positive Age Beliefs

1. For example, see Levy, B. R., Pilver, C., Chung, P. H., & Slade, M. D. (2014). Subliminal strengthening: Improving older individuals' physical function over time with an implicit-age-stereotype intervention. *Psychological Science, 25*, 2127–2135.

2. Donlon, M., & Levy, B. R. (2005). Re-vision of older television characters: Stereotype-awareness intervention. *Journal of Social Issues, 61*, 307–319.

3. Simons, D. J., Boot, W. R., Charness, N., Gathercole, S. E., Chabris, C. F., Hambrick, D. Z., Elizabeth, A. L., & Stine-Morrow, E. A. (2016). Do "brain training" programs work? *Psychological Science in Public Interest, 17*, 103–186; Federal Trade Commission(2016, January 5). Lumosity to pay $2 million to settle FTC deceptive advertising charges for its "brain training" program (2016, January 5). https://www.ftc.gov /news-events/press-releases/2016/01/lumosity-pay-2-million-settle-ftc -deceptive-advertising-charges.

4. Leblanc, R. (2019, March 12). Recycling beliefs vary between gener-

ations. The Balance Small Business. https://www.thebalancesmb.com/who-recycles-more-young-or-old-2877918.

5. Hall, D. (2018, February 4). Anatomy of a Super Bowl ad: Behind the scenes with E-Trade's ode to retirement. *AdAge*. https://adage.com/article/special-report-super-bowl/anatomy-a-super-bowl-ad-scenes-e-trade/312230.

6. Thomas, P. (2019). E-Trade profits jump, new users added. *The Wall Street Journal*. https://www.wsj.com/articles/e-trade-profit-jumps-new-users-added-11555536789.

Appendix 2: Ammunition to Debunk Negative Age Stereotypes

1. Gutchess, A. (2014). Plasticity of the aging brain: New directions in cognitive neuroscience. *Science, 346*, 579–582.

2. Roring, R. W., & Charness, N. (2007). A multilevel model analysis of expertise in chess across the life span. *Psychology and Aging, 22*, 291–299.

3. Mireles, D. E., & Charness, N. (2002). Computational explorations of the influence of structured knowledge on age-related cognitive decline. *Psychology and Aging, 17*, 245–259.

4. Park, D. C., Lodi-Smith, J., Drew, L., Haber, S., Hebrank, A., Bischof, G. N., & Aamodt, W. (2014). The impact of sustained engagement on cognitive function in older adults: The Synapse Project. *Psychological Science, 25*, 103–112.

5. Langa, K. M., Larson, E. B., Crimmins, E. M., Faul, J. D., Levine, D. A., Kabeto, M. U., & Weir, D. R. (2017). A comparison of the prevalence of dementia in the United States in 2000 and 2012. *JAMA Internal Medicine, 177*, 51–58.

6. 2021 Alzheimer's disease facts and figures. (2021). *Alzheimer's & Dementia, 17*, 391–460.

7. Wolters, F. J., Chibnik, L. B., Waziry, R., Anderson, R., Berr, C., Beiser, A., . . . & Hofman, A. (2020). Twenty-seven-year time trends in dementia incidence in Europe and the United States. *The Alzheimer Cohorts Consortium, 95*, e519–e531.

8. Levy, B. R., Hausdorff, J. M., Hencke, R., & Wei, J. Y. (2000). Reducing cardiovascular stress with positive self-stereotypes of aging. *The Journals of Gerontology, Series B: Psychological Sciences and Social Sciences, 55*, 205–213.

9. Levy, B. R., Slade, M. D., Kunkel, S. R., & Kasl, S. V. (2002). Longevity increased by positive self-perceptions of aging. *Journal of Personality and Social Psychology, 83*, 261–270.

10. Levy, B. R., & Myers, L. M. (2004). Preventive health behaviors influenced by self-perceptions of aging. *Preventive Medicine, 39,* 625–629.

11. Levy, B. R., Zonderman, A. B., Slade, M. D., & Ferrucci, L. (2012). Memory shaped by age stereotypes over time. *The Journals of Gerontology, Series B: Psychological Sciences and Social Sciences, 67,* 432–436.

12. Levy, B. R., Slade, M. D., Pietrzak, R. H., & Ferrucci, L. (2020). When culture influences genes: Positive age beliefs amplify the cognitive-aging benefit of *APOE ε2. The Journals of Gerontology, Series B: Psychological Sciences and Social Sciences, 75,* e198–e203.

13. Levy, B. R., Slade, M. D., Murphy, T. E., & Gill, T. M. (2012). Association between positive age stereotypes and recovery from disability in older persons. *JAMA, 308,* 1972–1973; World Health Organization. (2020). WHO guidelines on physical activity and sedentary behavior. Retrieved July 13, 2021, from https://www.who.int/publications/i /item/9789240015128.

14. Thomas, M. L., Kaufmann, C. N., Palmer, B. W., Depp, C. A., Martin, A. S., Glorioso, D. K., Thompson, W. K., & Jeste, D. V. (2016). Paradoxical trend for improvement in mental health with aging: A community-based study of 1,546 adults aged 21–100 years. *The Journal of Clinical Psychiatry, 77,* e1019–e1025. https://doi.org/10.4088 /JCP.16m10671; Fiske, A., Wetherell, J. L., & Gatz, M. (2009). Depression in older adults. *Annual Review of Clinical Psychology, 5,* 363–389; Villarroel, M. A., & Terlizzi, E. P. (2020). *Symptoms of depression among adults: United States, 2019.* National Center for Health Statistics Data Brief. https://www.cdc.gov/nchs/data/databriefs/db379-H.pdf.

15. Segal, D. L., Qualls, S. H., & Smyer, M. A. (2018). *Aging and mental health.* Hoboken, NJ: Wiley Blackwell.

16. Cuijpers, P., Karyotaki, E., Eckshtain, D., Ng, M. Y., Corteselli, K. A., Noma, H., Quero, S., & Weisz, J. R. (2020). Psychotherapy for depression across different age groups: A systematic review and meta-analysis. *JAMA Psychiatry, 77,* 694–702.

17. Age Smart Employer: Columbia Aging Center. (2021, May 24). The advantages of older workers. https://www.publichealth.columbia.edu /research/age-smart-employer/advantages-older-workers.

18. Börsch-Supan, A. (2013). Myths, scientific evidence and economic policy in an aging world. *The Journal of the Economics of Ageing, 1–2,* 3–15.

19. Conley, C. (2018). *Wisdom at work: The making of a modern elder.* New York: Currency.

20. Ibid. Loch, C., Sting, F., Bauer, N., & Mauermann, H. (2010). The globe:

How BMW is defusing the demographic time bomb. *Harvard Business Review*. https://hbr.org/2010/03/the-globe-how-bmw-is-defusing-the-demographic-time-bomb; Conley, C. (2018). *Wisdom at work: The making of a modern elder*. New York: Currency.

21. Frumkin, H., Fried, L., & Moody, R. (2012). Aging, climate change, and legacy thinking. *American Journal of Public Health, 102*, 1434–1438.

22. Konrath, S., Fuhrel-Forbis, A., Lou, A., & Brown, S. (2012) Motives for volunteering are associated with mortality risk in older adults. *Healthy Psychology, 31*, 87–96.

23. Benefactor. Sixty and over: Elders and philanthropic investments. https://benefactorgroup.com/sixty-and-over-elders-and-philanthropic-investments/.

24. Marketing to seniors and boomers: Ten things you need to know. Coming of Age. https://www.comingofage.com/marketing-to-seniors-and-boomers/.

25. LeBlanc, R. (2019, March 12). Recycling beliefs vary between generations. The Balance Small Business. https://www.thebalancesmb.com/who-recycles-more-young-or-old-2877918.

26. Mayr, U., & Freund, A. M. (2020). Do we become more prosocial as we age, and if so, why? *Current Directions in Psychological Science, 29*, 248–254; Lockwood, P. L., Abdurahman, A., Gabay, A. S., Drew, D., Tamm, M., Husain, M., & Apps, M. A. J. (2021). Aging increases prosocial motivation for effort. *Psychological Science, 32*, 668–681.

27. Betts, L. R., Taylor, C. P., Sekuler, A. B., & Bennett, P. J. (2005). Aging reduces center-surround antagonism in visual motion processing. *Neuron, 45*, 361–366.

28. Grossmann, I., Na, J., Varnum, M. E., Park, D. C., Kitayama, S., & Nisbett, R. E. (2010). Reasoning about social conflicts improves into old age. *Proceedings of the National Academy of Sciences, 107*, 7246–7250.

29. Pennebaker, J. W., & Stone, L. D. (2003). Words of wisdom: Language use over the life span. *Journal of Personality and Social Psychology, 85*, 291–301.

30. American Psychological Association. (2021). Memory and aging. https://www.apa.org/pi/aging/memory-and-aging.pdf; Nyberg, L., Maitland, S. B., Rönnlund, M., Bäckman, L., Dixon, R. A., Wahlin, Å., & Nilsson, L.-G. (2003). Selective adult age differences in an age-invariant multifactor model of declarative memory. *Psychology and Aging, 18*, 149–160.

31. Arkowitz, H., & Lilienfeld, S. O. (2012, November 1). Memory in old age can be bolstered. *Scientific American*. https://www.scientificamerican.com/article/memory-in-old-age-can-be-bolstered/.

32. Belleville, S., Gilbert, B., Fontaine, F., Gagnon, L., Ménard, É., & Gauthier, S. (2006). Improvement of episodic memory in persons with mild cognitive impairment and healthy older adults: Evidence from a cognitive intervention program. *Dementia and Geriatric Cognitive Disorders, 22,* 486–499.

33. Zimmermann, N., Netto, T. M., Amodeo, M. T., Ska, B., & Fonseca, R. P. (2014). Working memory training and poetry-based stimulation programs: Are there differences in cognitive outcome in healthy older adults? *NeuroRehabilitation, 35,* 159–170.

34. Levy, B. R., Zonderman, A. B., Slade, M. D., & Ferrucci, L. (2012). Memory shaped by age stereotypes over time. *The Journals of Gerontology, Series B: Psychological Sciences and Social Sciences, 67,* 432–436.

35. Levy, B. R., & Leifheit-Limson, E. (2009). The stereotype-matching effect: Greater influence on functioning when age stereotypes correspond to outcomes. *Psychology and Aging, 24,* 230–233.

36. Marottoli, R. A., & Coughlin, J. F. (2011). Walking the tightrope: Developing a systems approach to balance safety and mobility for an aging society. *Journal of Aging & Social Policy, 23,* 372–383; Leefeldt, E., & Danise, A. (2021, March 16). Senior drivers are safer than previously thought. *Forbes.* https://med.fsu.edu/sites/default/files/news -publications/print/Senior%20Drivers%20Are%20Safer%20Than%20 Previously%20Thought%20-%20Forbes%20Advisor.pdf; American Occupational Therapy Association. Myths and realities about older drivers. https://www.aota.org/Practice/Productive-Aging/Driving/Clients /Concern/Myths.aspx.

37. Bergal, J. (2016, December 15). Should older drivers face special restrictions? Pew Charitable Trusts. https://www.pewtrusts.org/en/res earch-and-analysis/blogs/stateline/2016/12/15/should-older-drivers -face-special-restrictions.

38. Tortorello, M. (2017, June 1). How seniors are driving safer, driving longer. *Consumer Reports.* https://www.consumerreports.org/elderly -driving/how-seniors-are-driving-safer-driving-longer/.

39. Bunis, D. (2018, May 3). Two-thirds of older adults are interested in sex, poll says. AARP. https://www.aarp.org/health/healthy-living/info -2018/older-sex-sexual-health-survey.html.

40. Kalra, G., Subramanyam, A., & Pinto, C. (2011). Sexuality: Desire, activity and intimacy in the elderly. *Indian Journal of Psychiatry, 53,* 300–306.

41. Azoulay, P., Jones, B. F., Kim, J. D., & Miranda, J. (2018, April). Age

and high-growth entrepreneurship. National Bureau of Economic Research Working Paper No. w24489. Available at SSRN: https://ssrn.com/abstract=3158929.

42. Rietzschel, E. F., Zacher, H., & Stroebe, W. (2016). A lifespan perspective on creativity and innovation at work. *Work, Aging and Retirement, 2,* 105–129.

43. American Psychological Association Office on Aging. (2017). Older adults' health and age-related changes: Reality versus myth. https://www.apa.org/pi/aging/resources/guides/myth-reality.pdf.

44. Swayne, M. (2019). Consider older adults as "leaders in innovation." Futurity. https://www.futurity.org/leaders-in-innovation-aging-older-adults-2094222-2/.

45. AARP. (2019, December). 2020 Tech and the 50+ survey. https://www.aarp.org/content/dam/aarp/research/surveys_statistics/technology/2019/2020-tech-trends-survey.doi.10.26419-2Fres.00329.001.pdf.

46. Czaja, S. J., Boot, W. R., Charness, N., & Rogers, W. A. (2019). *Designing for older adults: Principles and creative human factors approaches.* Boca Raton, FL: CRC Press.

47. Pontin, J. (2013, August 13). Seven over seventy. *MIT Technology Review.* https://www.technologyreview.com/2013/08/21/176715/seven-over-70-4/.

48. National Institute on Aging. (2019, January 17). Quit smoking for older adults. https://www.nia.nih.gov/health/quitting-smoking-older-adults; Yassine, H. N., Marchetti, C. M., Krishnan, R. K., Vrobel, T. R., Gonzalez, F., & Kirwan, J. P. (2009). Effects of exercise and caloric restriction on insulin resistance and cardiometabolic risk factors in older obese adults—a randomized clinical trial. *The Journals of Gerontology, Series A: Biomedical Sciences and Medical Sciences, 64,* 90–95.

49. Ibid.

50. Hardy, S. E., & Gill, T. M. (2004). Recovery from disability among community-dwelling older persons. *JAMA, 291,* 1596–1602; Levy, B. R., Slade, M. D., Murphy, T. E., & Gill, T. M. (2012). Association between positive age stereotypes and recovery from disability in older persons. *JAMA, 308,* 1972–1973.

Appendix 3: A Call to End Structural Ageism

1. Irving, P. (2014). *The upside of aging: How long life is changing the world of health, work, innovation, policy, and purpose.* Hoboken, NJ: Wiley, p. xxi.

2. Chang, E., Kannoth, K., Levy, S., Wang, S., Lee, J. E., & Levy, B. R. (2020). Global reach of ageism on older persons' health: A systematic review. *PLOS ONE.* https://doi.org/10.1371/journal.pone.0220857.

3. Young, P. L., & Olsen, L. (2010). *The healthcare imperative: Lowering costs and improving outcomes.* National Academy of Sciences. https://www.ncbi.nlm.nih.gov/books/NBK53906/.

4. Tinetti, M. E., Costello, D. M., Naik, A. D., Davenport, C., Hernandez-Bigos, K., . . . Dindo, L. (2021). Outcome goals and health care preferences of older adults with multiple chronic conditions. *JAMA Network Open, 4*(3), e211271. https://doi.org/10.1001/jamanetwork open.2021.1271; Tinetti, M. E., Naik, A., & Dindo, L. (2018). *Conversation guide and manual for identifying patients' health priorities.* Patient Priorities Care. https://patientprioritiescare.org/wp-content /uploads/2018/11/Conversation-Guide-and-Manual-for-Identifying -Patients27-Health-Priorities.pdf.

5. Southerland, L. T., Lo, A. X., Biese, K., Arendts, G., Banerjee, J., Hwang, U., . . . & Carpenter, C. R. (2020). Concepts in practice: Geriatric emergency departments. *Annals of Emergency Medicine, 75,* 162–170; Hwang, U., Dresden, S. M., Vargas-Torres, C., Kang, R., Garrido, M. M., Loo, G., . . . & Structural Enhancement Investigators. (2021). Association of a Geriatric Emergency Department Innovation Program with cost outcomes among Medicare beneficiaries. *JAMA Network Open, 4,* e2037334–e2037334.

6. The median annual salary for geriatricians is $189,879, according to Salary.com. That's less than half of what an average orthopedic surgeon or cardiologist earns. Geriatricians earn less because Medicare is their primary payer and offers historically lower reimbursement rates than commercial insurance. Hafner, K. (2016, January 25). As population ages, where are the geriatricians? *The New York Times.* https://www .nytimes.com/2016/01/26/health/where-are-the-geriatricians.html; Castellucci, M. (2018, February 27). Geriatrics still failing to attract new doctors. *Modern Healthcare.* https://www.modernhealthcare.com /article/20180227/NEWS/180229926/geriatrics-still-failing-to-attract -new-doctors.

7. Massachusetts Care Planning Council. (2013). About geriatric health care. https://www.caremassachusetts.org/services_members/09_about _geriatric_medical_care.htm; Hafner, K. (2016, January 25). As population ages, where are the geriatricians? *The New York Times.* https://www .nytimes.com/2016/01/26/health/where-are-the-geriatricians.html.

8. McGinnis, S. L., & Moore, J. (2006, April–June). The impact of the aging population on the health workforce in the United States—summary of key findings. *Cahiers de Sociologie et de Démographie Médicales, 46,* 193–220; Bardach, S. H., & Rowles, G. D. (2012). Geriatric education in the health professions: Are we making progress? *The Gerontologist, 52,* 607–618.

9. Fulmer, T., Reuben, D. B., Auerbach, J., Fick, D., Galambos, C., & Johnson, K. S. (2021). Actualizing better health and health care for older adults. *Health Affairs, 40,* 219–225; Institute of Medicine, Committee on the Future Health Care Workforce for Older Americans (2008). *Retooling for an aging America: Building the health care workforce.* Washington, DC: National Academies Press.

10. Seegert, L. (2019, June 26). Doctors are ageist—and it's harming older patients. NBC News. https://www.nbcnews.com/think/opinion/doctors -are-ageist-it-s-harming-older-patients-ncna1022286. Makris, U. E., Higashi, R. T., Marks, E. G., Fraenkel, L., Sale, J. E., Gill, T. M., & Reid, M. C. (2015). Ageism, negative attitudes, and competing co-morbidities— why older adults may not seek care for restricting back pain: A qualitative study. *BMC Geriatrics, 15,* 39. https://doi.org/10.1186/s12877-015 -0042-z.

11. Bodner, E., Palgi, Y., & Wyman, M. (2018). Ageism in mental health assessment and treatment of older adults. In L. Ayalon & C. Tesch-Römer (Eds.), *Contemporary perspectives on ageism.* New York: Springer; Bouman, W. P., & Arcelus, J. (2001). Are psychiatrists guilty of "ageism" when it comes to taking a sexual history? *International Journal of Geriatric Psychiatry, 16,* 27–31; Lileston, R. (2017, September 28). STD rates keep rising for older adults. AARP. https://www.aarp.org/health/conditions -treatments/info-2017/std-exposure-rises-older-adults-fd.html.

12. Johnson, S. R. (2016, October 8). Payment headaches hinder progress on mental health access. *Modern Healthcare.* https://www.modernhealth care.com/article/20161008/MAGAZINE/310089981/payment-head aches-hinder-progress-on-mental-health-access.

13. Chibanda, D., Weiss, H. A., Verhey, R., Simms, V., Munjoma, R., Rusakaniko, S., . . . & Araya, R. (2016). Effect of a primary care–based psychological intervention on symptoms of common mental disorders in Zimbabwe: A randomized clinical trial. *JAMA, 316,* 2618–2626.

14. Aging in place. (2021, June). The facts behind senior hunger. https:// aginginplace.org/the-facts-behind-senior-hunger/; America's Health Ratings. (2021). Poverty 65+. https://www.americashealthrankings.org

/explore/senior/measure/poverty_sr/state/ALL; Nagourney, A. (2016, May 31). Old and on the street: The graying of America's homeless. *The New York Times*. https://www.nytimes.com/2016/05/31/us/americas-ag ing-homeless-old-and-on-the-street.html.

15. U.S. Department of Labor. Legal highlight: The Civil Rights Act of 1964. https://www.dol.gov/agencies/oasam/civil-rights-center/statutes /civil-rights-act-of-1964.

16. Brown, B. (2018, April 24). Bethany Brown discusses human rights violations in US nursing homes. Yale Law School. https://law.yale.edu /yls-today/news/bethany-brown-discusses-human-rights-violations-us -nursing-homes; Human Rights Watch. (2018). "They want docile": How nursing homes in the United States overmedicate people with de- mentia. https://www.hrw.org/report/2018/02/05/they-want-docile/how -nursing-homes-united-states-overmedicate-people-dementia#_ftn64); Ray, W. A., Federspiel, C. F., & Schaffner, W. (1980). A study of anti- psychotic drug use in nursing homes: Epidemiologic evidence suggesting misuse. *American Journal of Public Health, 70,* 485–491; Thomas, K., Gebeloff, R., & Silver-Greenberg, J. (2021, September 11). Phony di- agnoses hide high rates of drugging at nursing homes. *The New York Times.* https://www.nytimes.com/2021/09/11/health/nursing-homes -schizophrenia-antipsychotics.html

17. Yon, Y., Mikton, C. R., Gassoumis, Z. D., & Wilber, K. H. (2017). Elder abuse prevalence in community settings: A systematic review and meta- analysis. *Lancet Global Health, 5*(2), e147–e156. https://doi.org/10.1016 /s2214-109x(17)30006-2; World Health Organization. (2021, June 15). Elder abuse. https://www.who.int/news-room/fact-sheets/detail/elder -abuse.

18. Chang, E. S., Monin, J. K., Zelterman, D., & Levy, B. R. (2021). Impact of structural ageism on greater violence against older persons: A cross- national study of 56 countries. *BMJ Open, 11*(5), e042580. https://doi .org/10.1136/bmjopen-2020-042580.

19. Organization of American States Department of International Law. (2015, June 15). Inter-American Convention on Protecting the Human Rights of Older Persons. http://www.oas.org/en/sla/dil/inter_american _treaties_A-70_human_rights_older_persons_signatories.asp.

20. National Center for State Courts. (2020, September 30). Mandatory judicial retirement. https://www.ncsc.org/information-and-resources /trending-topics/trending-topics-landing-pg/mandatory-judicial -retirement; ElderLawAnswers. (2019, March 12). Called for jury

duty? You may be exempt based on your age. https://www.elderlawan
swers.com/called-for-jury-duty-you-may-be-excused-based-on-your-age
-15650.

21. McGuire, S. (2020). Growing up and growing older: Books for young
readers. Lincoln Memorial University. https://library.lmunet.edu/book
list.

22. AARP Foundation. (2021). Experience Corps: Research studies. https://
www.aarp.org/experience-corps/our-impact/experience-corps-research
-studies/.

23. The Gerontological Society of America. Age-Friendly University (AFU)
global network. https://www.geron.org/programs-services/education
-center/age-friendly-university-afu-global-network.

24. Lieberman, A. (2018, February 27). UN increases retirement ages for
staffers to 65 years. Devex. https://www.devex.com/news/un-increases
-retirement-ages-for-staffers-to-65-years-92194.

25. Kita, J. (2019, December 30). Workplace age discrimination still flour-
ishes in America. AARP. https://www.aarp.org/work/working-at-50
-plus/info-2019/age-discrimination-in-america.html.

26. PayChex. (2016). Potential benefits of multigenerational workforce.
https://humanresources.report/Resources/Whitepapers/c79541ca
-d201-4cd1-a334-50642d8e2de2_whitepaper-potential-benefits-of-a
-multigenerational-workforce.pdf.

27. Centre for Ageing Better. (2021, January 7). Age-positive image library
launched to tackle negative stereotypes of later life. https://www.ageing
-better.org.uk/news/age-positive-image-library-launched,

28. Dan, A. (2016, September 13). Is ageism the ugliest "ism" on Madison Ave? Forbes. https://www.forbes.com/sites/avidan/2016/09/13/is
-ageism-the-ugliest-ism-on-madison-avenue/?sh=695108ae557c.

29. Sperling, N. (2020, September 8). Academy explains diversity rules
for best picture Oscar. The New York Times. https://www.nytimes
.com/2020/09/08/movies/oscars-diversity-rules-best-picture.html?.

30. Donlon, M., & Levy, B. R. (2005). Re-vision of older television characters: Stereotype-awareness intervention. Journal of Social Issues, 61,
307–319; Robinson, T., Callister, M., Magoffin, D., & Moore, J. (2007).
The portrayal of older characters in Disney animated films. Journal of
Aging Studies, 21, 203–213; Kessler, E., Rakoczy, K., & Staudinger,
U. M. (2004). The portrayal of older people in prime time television
series: The match with gerontological evidence. Ageing and Society, 24,
531–552.

31. Handy, B. (2016, May 3). Inside Amy Schumer: An oral history of Amy Schumer's "Last Fuckable Day" sketch. *Vanity Fair.* https://www.vanityfair.com/hollywood/2016/05/amy-schumer-last-fuckable-day; Comedy Central. (2015, April 22). *Inside Amy Schumer*—Last F**kable Day. [Video]. YouTube. https://www.youtube.com/watch?v=XPpsI8mWKmg; *Vogue.* (2019, May 3). Madonna on motherhood and fighting ageism: "I'm being punished for turning 60." https://www.vogue.co.uk/article/madonna-on-ageing-and-motherhood; de Souza, A. (2015, September 23). Ageism exists in Hollywood, says Robert De Niro. *The Straits Times.* https://www.straitstimes.com/lifestyle/entertainment/ageism-exists-in-hollywood-says-robert-de-niro-72.

32. Peterson, S. (2018, April 4). Ageism: The issue never gets old. Gamesindustry.biz. https://www.gamesindustry.biz/articles/2018-04-04-ageism-in-games-the-issue-never-gets-old; Dunn, P. (2021, April 22). "Just Die Already" to launch on all platforms next month. GBATEMP. https://gbatemp.net/threads/just-die-already-to-launch-on-all-platforms-next-month.587365/.

33. Changing the Narrative. (2021, February 8). Changing the Narrative's age-positive birthday card campaign includes artist from Aurora. Your Hub. https://yourhub.denverpost.com/blog/2021/02/changing-the-narratives-age-positive-birthday-card-campaign-includes-artist-from-aurora/273871/; Changing the Narrative. (2020, October 1). Anti-ageist birthday cards across cultures. https://changingthenarrativeco.org/2020/10/01/anti-ageist-birthday-cards-uk/.

34. Gabbatt, A. (2019, March 28). Facebook charged with housing discrimination in targeted ads. *The Guardian.* https://www.theguardian.com/technology/2019/mar/28/facebook-ads-housing-discrimination-charges-us-government-hud; The Associated Press. (2020, July 2). Lawsuit accuses property managers of ageist ads. Finance & Commerce. https://finance-commerce.com/2020/07/lawsuit-accuses-property-managers-of-ageist-ads/; Kofman, A., & Tobin, A. (2019, December 13). Facebook ads can still discriminate against women and older workers, despite a civil rights settlement. ProPublica. https://www.propublica.org/article/facebook-ads-can-still-discriminate-against-women-and-older-workers-despite-a-civil-rights-settlement.

35. Jimenez-Sotomayor, M. R., Gomez-Moreno, C., & Soto-Perez-de-Celis, E. (2020). Coronavirus, ageism, and Twitter: An evaluation of tweets about older adults and COVID-19. *Journal of the American Geriatrics Society, 68,* 1661–1665.

36. Facebook Community Standards. Objectionable content: Hate speech. Retrieved March 14, 2021, from https://www.facebook.com/communitystandards/.

37. Sipocz, D., Freeman, J. D., & Elton, J. (2021). "A toxic trend?": Generational conflict and connectivity in Twitter discourse under the #Boomer-Remover hashtag. *Gerontologist, 61,* 166–175.

38. Levy, B. R., Chang, E.-S., Lowe, S., Provolo, N., & Slade, M. D. (2021). Impact of media-based negative and positive age stereotypes on older individuals' mental health during the COVID-19 pandemic. *The Journals of Gerontology, Series B: Psychological Sciences and Social Sciences.* https://doi.org/10.1093/geronb/gbab085.

39. Leardi, J. (2015, June 10). Turning the tide on the "silver tsunami." Changing Aging with Dr. Bill Thomas. https://changingaging.org/blog/turning-the-tide-on-the-silver-tsunami/.

40. Older Adults Technology Services. (2021). Aging connected: Exposing the hidden connectivity crisis for older adults. https://oats.org/wp-content/uploads/2021/01/Aging-Connected-Exposing-the-Hidden-Connectivity-Crisis-for-Older-Adults.pdf.

41. National Aging and Disability Transportation Center. (2021). Older adults and transportation. https://www.nadtc.org/about/transportation-aging-disability/unique-issues-related-to-older-adults-and-transportation/.

42. The Anti-Ageism Taskforce at the International Longevity Center. (2006). *Ageism in America.* New York: International Longevity Center-USA, p. 41.

43. Nanna, M. G., Chen, S. T., Nelson, A. J., Navar, A. M., & Peterson, E. D. (2020). Representation of older adults in cardiovascular disease trials since the inclusion across the lifespan policy. *JAMA Internal Medicine, 180,* 1531–1533; Gopalakrishna, P. (2020, September 30). New research shows older adults are still often excluded from clinical trials. STAT. https://www.statnews.com/2020/09/30/age-disparities-clinical-trials-covid19/.

44. Albone, R., Beales, S., & Mihnovits, A. (2014, January). Older people count: Making data fit for purpose. Global AgeWatch. https://www.helpage.org/silo/files/older-people-count-making-data-fit-for-purpose.pdf.

45. Center on Budget and Policy Priorities. (2020, April 9). Policy basics: Where do our federal tax dollars go? https://www.cbpp.org/research/federal-budget/where-do-our-federal-tax-dollars-go.

INDEX